P9-DWQ-506

The Genetic Inferno
Inside the Seven Deadly Sins

What makes us react or feel the way we do? If you have ever asked yourself this question, then let gifted writer John Medina take you on a tour of the fascinations and frustrations involved in the quest to understand the biological basis of human behavior.

By describing the gap that exists between a human behavior and a human gene, this fascinating book seeks both to clarify and debunk ideas about the genetic roots of behavior, from the genes of divorce to the tendency to eat chocolate. Using Dante's *The Divine Comedy* as an organizing framework, *The Genetic Inferno* explains each of the Seven Deadly Sins, but in terms of modern understanding of genes and brains. Written by a practising research scientist, this book is not for biologists, but for literature majors, business people, parents, and anyone interested in how our genes work to make us behave the way we do.

JOHN MEDINA is a molecular biologist formerly on the faculty of the Department of Bioengineering at the University of Washington School of Medicine and now Founding Director of the Center for Mind, Brain and Learning. He is also a consultant and regular columnist for an American psychiatric organization on the genetics and neurobiology of human behavior. In the course of his research career, which includes the isolation and characterization of genes involved in cardiovascular development, Dr Medina became concerned with the public communication of biological sciences to both lay and professional medical audiences. He was recently recognized as the Merrill Dow/Continuing Medical Education Teacher of the Year.

Bosch's Seven Deadly Sins © *The Prado National Museum, Madrid, Spain.*
(Reproduced in full here, and in sections for each chapter opening page.)

The **Genetic Inferno**
Inside the Seven Deadly Sins

Written and illustrated by John J. Medina
University of Washington School of Medicine, Seattle

CAMBRIDGE
UNIVERSITY PRESS

PUBLISHED BY THE PRESS SYNDICATE OF THE UNIVERSITY OF CAMBRIDGE
The Pitt Building, Trumpington Street, Cambridge, United Kingdom

CAMBRIDGE UNIVERSITY PRESS
The Edinburgh Building, Cambridge CB2 2RU, UK
40 West 20th Street, New York, NY 10011-4211, USA
10 Stamford Road, Oakleigh, VIC 3166, Australia
Ruiz de Alarcón 13,28014 Madrid, Spain
Dock House, The Waterfront, Cape Town 8001, South Africa
http://www.cambridge.org

© John Medina 2000

This book is in copyright. Subject to statutory exception
and to the provisions of relevant collective licensing agreements,
no production of any part may take place without
the written permission of Cambridge University Press.

First published 2000

Printed in the United Kingdom at the University Press, Cambridge

Typeface Hollander 9/13pt System QuarkXPress® [w21]

A catalogue record for this book is available from the British Library

Library of Congress Cataloguing in Publication data

Medina, John, 1956-
    The genetic inferno : inside the seven deadly sins / written and illustrated
by John J. Medina.
        p. cm.
    Includes bibliographical references.
    ISBN 0 521 64064 4 (hardcover)
    1. Genetic psychology. 2. Behavior genetics. 3. Deadly sins. I. Title.
BF711 .M43 2000
155.7–dc21   99-086430

    ISBN 0 521 64064 4 hardback

Contents

Introduction

"Yours am I, sacred Muses! To you I pray.
Here let dead poetry rise once more to life,
and here let sweet Calliope rise and play

some far accompaniment in that high strain
whose power the wretched Pierides once felt
so terribly they dared not hope again."

-Canto I, The Purgatorio

The first time I ever heard my newborn son cry, I also heard my deceased mother's voice. Here was this little pickle of a baby, all wrinkly and salty and crying, kind of like my wife and I at that moment – and here was this memory of a middle-aged woman exclaiming to me, "You are *just* like your dad!" So common to new fathers in any century, the birth was a confusing, delirious, exhilarating moment to be alive.

Though I smiled pleasantly remembering my mother's voice, her genetic comments were actually founded in anger. I was five years old when she said those words, spoken just after my grade-school chum Scottie and I had seen our first big-city aquarium. I remember being very excited. After school, I immediately dashed over to Scottie's house and decided right there to simulate some of the interesting stuff we had experienced that day. We even discussed a starting premise, taken from a sudden revelation Scottie had about some trout in his parents' freezer. Specifically, he postulated that frozen fish were the biological equivalent of sleeping fish. If we could find a way to wake them up, we'd have our own big-city aquarium! In this spirit of excited behavioral inquiry, he took four frozen trout from the freezer and plopped them into his family's tropical fish aquarium. Then we pressed our noses against the tank, looking eagerly as the quartet of German Browns sank like rocks to the bottom. We were sorely disappointed of course; the only activity was a passing and somewhat bewildered angelfish, who picked at the fins of the frozen invaders. These actions were not what my mother's anger was about, however, though she was referencing my father's similar penchant for tinkering with the way things worked. This tinkering tendency must have rubbed off, because I am now a research geneticist (a developmental molecular biologist, often called in the popular press a "genetic engineer") just as interested in the way biological things work, still rejecting and accepting hypotheses about physical phenomena. I suppose that same spirit has always been with me, certainly by the time I was five, gazing at the failure of the cold fish experiment in the aquarium.

Scottie and I ignored the negative data, a trait I now know to be common among scientists, kept faith in our hypothesis, and decided to do another experiment. Was the aquarium too small to wake up the sleeping trout? If the fish were given a larger venue, perhaps then they would be roused. The largest encloseable area we could think of was the station wagon (owned by Scottie's parents) sitting out in the driveway. And then Scottie suddenly laughed, exclaiming, "I have a *great* idea!" and grabbed the fish out of the aquarium. He ran as fast as he could to the family car, and threw the now-thawing burdens into the front seat. Scottie then rolled up all the windows except one, left open just a crack to accommodate the next technological step of the experiment.

It was this step that got us into trouble, powerfully reminding my mother of her youngest son's paternity. With my help, Scottie ran back to the house and grabbed the garden hose. We unraveled it together, Scottie dragging it over to the car, inserting it into the almost-shut window. He instructed me to turn the water on full blast, which I did enthusiastically. . . .

For awhile, the experiment actually seemed to work. As the car flooded with water, we could see the fish appearing to float a little. It wasn't quite like the giant commercial aquarium – and the fish weren't really floating – but it was satisfying indeed for a couple of curious five year olds. Proud of his accomplishment, Scottie rushed back into his house to tell his dad what had happened, and I waited by the faucet with a triumphant smile on my face.

I can only relate that in the next minute my grin left me and my ears began to ache, the pain foreshadowing what would soon be felt by my rear-end. An ex-marine, Scottie's father let out a scream that could have awakened the dead, rushed out of the house, and yanked the hose out of his mostly ruined car. I then heard the retired military man make two further genetic references, first about the sexual history of Scottie's mother, then about my mother's, with me now paralyzed in fear against the side of the house. It was after Scottie's father dumped my trembling body on the doorstep of my own home that I heard my mother's declaration about my paternally derived tendencies for mischief. Mom said it was the biggest spanking she ever gave me.

That was many years ago, of course, with the hot terror of the moment cooled to a warm, funny memory. As I sat there reminiscing about my first real scientific adventure, I was suddenly brought back to the birthing room by more soft cries – even a yawn! – of our newborn. I began to wonder aloud how many of my own cars would be ruined before my new son heard the cries of *his* child. I felt my mother's voice fade to the background as I heard my wife laugh, and I gave the baby to her, teary, trying unsuccessfully not to think about the wonder of it all.

The purpose of this book

I don't think any amount of research experience prepares you for the birth of your own child. I have been cloning human genes for many years now, virtually since it was possible. Yet in the delivery room, viewing the combined results of the greatest genetic engineering project possible, I mostly just blubbered and cried. So many colliding feelings. So little time to sort them all. How could such a little guy, a near eight-pounder named Joshua, produce such an incredible weight of – what? – joy? pride? fear? all of the above? The feelings were almost too intense to form questions. What kind of mistakes would his mother and I make raising him? What kind of magnifying glass does Joshua have that could so powerfully focus all his Daddy's emotion into one tiny, wiggly little space? I am reminded of a verse from *The Purgatorio*, one of sections that comprise Dante Alighieri's *The Divine Comedy*:

> *Thus, you may understand that love alone*
> *Is the true seed of every merit in you,*
> *And of all acts for which you must atone.*

The birth of a child is powerful not for just the emotions that the experience elicits. Tucked in the back of your mind, like a collection of ticking bombs, are all the worries, excitements and anticipations about what the child will be like. In recalling my mother's words, I wondered if this newborn would have his Daddy's curiously mischievous streak, taken from grandfather's genetic armory. I wondered if he would have my wife's gift for music, or her kindness, or her beauty. What types of things are heritable anyway? Would his disposition turn out to be just as genetic as his currently cobalt blue eyes?

In an indirect way, the book you have in your hand attempts to address the worries every new mother and father face as they contemplate the nature of their newborn child. It is a tour, really, through some of the most captivating and frustrating biological research that exists: the quest to understand the biological basis of human behaviors. Specifically, I want to talk about seven different human behaviors, all except one falling under the category of "emotion". I want to talk about them at the most intimate level possible, starting at the level of the brain, and moving to the level of the gene.

Don't let the words *biological basis* scare you away. Though this book is about research science, it is not written for research professionals, nor is it meant to be a formal discussion of the broad topic of human behavior. Rather, it is written for literature majors, political science types, business people, anyone who has an interest in human beings but who is not a working biochemist. Moreover, these pages are not meant to be exhaustive, but to provide simply a glimpse into the mysterious bridge between genes and human behaviors. To underscore this fact, I use as an organizing principle behaviors not extracted from a biological point of view, but from a literary perspective. Specifically, I exploit chapters from *The Divine Comedy*, written by the Dante from whom I just quoted, focusing on the biology of the so-called Seven Deadly Sins. You are probably familiar with these sins, and not just because you have read about them; a powerful organizing feature of human behavior in medieval times, The Seven Deadly Sins still resonate with the people of the twenty-first century. There are chapters on lust and sloth and wrath and gluttony, for example, though these days we are more likely to say words like sexual arousal and circadian rhythms and fight-or-flight responses and appetite control.

Admittedly, there is a lot to chew on here. To make this text easier to swallow, I use plenty of metaphors and analogies taken from the pages of *The Divine Comedy*, as well as anecdotes from thirteenth and fourteenth century medieval life. As I was a graphics artist before I was a scientist, I have also included plenty of illustrations to get us through parts of the biology that might seem a bit unfamiliar.

When appropriate, I also mention conversations and lectures from various scholars and academic professionals (they go unnamed to protect their privacy and my sometimes lapse of memory) whose specialties were not the sciences, but the humanities. In fact, each chapter starts with a classroom lecture given by an imaginary medievalist professor. To introduce the subject of a given biological exploration, the medievalist describes what happens to Dante at the appropriate level of the afterlife mentioned in *The Divine Comedy*. I use the word imaginary here because the dialogue represents the mosaic distillation of many academics and numerous lectures I consulted while putting this book together.

At the end of each chapter I include a subject that was truly brought home to me when I heard our newborn cry. As we traverse the genes and biochemicals involved in certain behaviors, you may at some point ask yourself, "Does Medina think we are simply a vast array of sophisticated chemical processes?" If that's what crosses your mind, you are in familiar company. As my wailing Joshua looked up at me with eyes as blank and as full as a deep ocean, I sensed the uneasy truce science has made with issues of human identity. The issues of "who we are" were a part of Dante's time, too, though couched in different, primarily religious, forms. I could not write a book describing human behaviors without giving at least a nod to one part of this issue: the history of what has commonly been called the "mind/brain" dilemma. This is the idea that contemplates whether a mind

– and thus a human – is more than the sum of its neurological parts. We will start with Greek notions about a soul, work through Descartes *cogito ergo sum* and end with a controversial idea some neuroscientists are calling emergentism.

Finally, each chapter will conclude with a discussion of how mind/brain ideas impact on the concept of the emotion being considered. As we shall see on p. 8, we are going to use as a working model the idea that an emotion is made of two parts: an embedded neural system interacting at some level with our consciousness. That will take some explaining, for, as you may know, the modern notion of consciousness has roots deeply connected to mind/brain discussions. As we shall also note later, no one really knows what consciousness is, though one can easily list some of the ingredients of which it must be composed. And that's exactly how we'll close each chapter, discussing one of the many ingredients that make up the notion of consciousness, and showing how ingredients may inform the biology of the emotion being considered in the individual chapter. That should put to rest any questions about how well the uneasy truce between science and human identity issues is holding up.

Taken together then, each chapter can be divided into five parts. If it gets confusing, feel free to recheck this outline:

1. A quick word with Dante, describing the section of *The Purgatorio* he is currently visiting.
2. The biology of the emotion under study (starting with cells, moving to genes).
3. A segment of history regarding mind/brain dilemmas.
4. An ingredient of consciousness, and its interface with mind/brain issues.
5. A parting thought about how the ingredient informs the discussion of the emotion in question.

So let's get started

As I stated above, I make no attempt to be thorough in this book. In fact, the point I wish to leave you with is this: in discussing DNA and human motivation, it is not *possible* to be thorough, because there is too much that is unknown. The gap between a gene and a behavior is so vast that the best explanation might invoke not medieval, but cosmological, metaphors. Even when gene functions have been responsibly described, one does not find single genes causing identifiable behaviors, but collective efforts of many genes appearing to make certain neurological impressions.

Even if that is all I get across here, demonstrating that there is a distance between a behavior and a gene will be well worth the effort, and for a surprising reason. There is some terrific research going on in the field of human behavioral genetics. Blossoming disciplines such as neurological psychiatry point not only to our increased knowledge of natural processes, but our understanding of what happens when things go wrong. Obscuring these great strides by overinterpretation – or the close sin of oversimplification – denigrates not only the data, but the dedicated individuals who are discovering them.

Before we get into the main body of our discussion, however, we need to describe some important biological background in both neuroscience and molecular biology. Only with that information in mind will we be ready to borrow some fourteenth century text to define the borders of some twenty-first century research. And we will be equipped to put some research ears to the infinitesimal chasm of human behavior, determining if the innocent cries of babies and the exasperated cries of mothers create an echo large enough to cross over from the fourteenth century. Or small enough to be heard in a test tube.

CHAPTER ONE

The Power of Physics Envy

"As I have told you, I was sent to show
the way his soul must take for its salvation;
and there is none but this by which I go.

I have shown him the guilty people. Now I mean
to lead him through the spirits in your keeping,
to show him those whose suffering makes them clean."

-Canto I, The Purgatorio

How's this for a battlefield cure?

> Obtain a half pound of grease from a wild boar and a tame one and
> the same quantity of bear fat. Gather a goodly portion of earthworms,
> place everything into a pot, seal it and cook lightly. Take a quantity
> of moss which has grown from the skull of a hanged man and press
> it into the shape of four walnuts. Add a little wine and mix all. You
> have created Unguentum Armarium, the ointment of war.

Sound tasty? This quote is from a medieval text on the preferred treatment
of battlefield wounds that are caused by weapons of war, such as swords.
In medieval days, it could be very difficult to tell which was worse, the
wound or cures such as these. The placement of such an ointment directly
on a laceration probably would have created a life-threatening situation
even if none had existed! Surprisingly, the reality is that this rather eclectic
mixture of elements probably did little harm to a medieval warrior. The
reason comes from the further directions in the text: the ointment was
not to be placed onto the wound but rather onto the weapon that made
the wound. When that occurred, the patient was guaranteed to experience
the wondrous healing powers of the *Unguentum Armarium.*

From a modern-day perspective, the last two paragraphs seem like a medical nightmare; there is so much wrong physiologically that it is difficult
to know where to begin labeling the errors. The most obvious sin here is
oversimplification, based on inadequate data, in turn based on faulty
assumptions. There might be a certain logic to thinking that the wound-
causing weapon might also contain the source of healing (the idea is *very*
much in vogue in modern psychiatry: the therapist might tell you to con-
front a person who is the source of all your troubles in order to initiate
healing, for example). But the ointment of war doesn't necessarily make
biological sense to groups of wounded cells, inflammatory responses or
immune systems. To correctly treat a laceration, one would need a background in the physiology of human wound response, and a working knowledge of germ theory. As of this writing, the majority of these data aren't
even one hundred years old.

In this introduction, I want to explore this idea that a solid background
is necessary to keep us from oversimplifying important biological processes. I don't bring this up to satirize old medieval texts, but to address
some current misperceptions about genes and human behaviors. I don't
know how many times I've thrown down a newspaper after reading things
like, "scientists have isolated the genes responsible for adultery" or an

article describing the, "DNA behind the desire to eat chocolate", or that there are actually chromosomes responsible for the predilection to vote Republican. The attitude that, "if you have the gene, then you have the tendency" occurs with such frequency that many of us who wear lab coats have quit reading the popular press. From a researcher's perspective, most of these headlines carry no more scientific integrity than an ointment made from a hanged man's skull-garden.

The problem is a dearth of background, and the temptation to simplify because of the honest desire both to understand and to avoid complicated issues. I want to talk about both of these desires in this chapter, and, to accomplish this, the chapter is divided into three parts. We will first consider how to scientifically study aspects of human behavior, some pitfalls, some unproductive assumptions, even a few controversies. Then we will proceed to the biology itself, describing some of the brain, neural and genetic interactions necessary to understand the data described in the rest of this book. Finally, we will talk a bit more about Dante and his *Divine Comedy*, contrasting how he organized human behavior with what modern biologists observe.

Let's begin our discussion not with biology but with a far older scientific discipline, physics, and the elegant math that undergirds it. We start with physics and not biology for reasons of simplicity. Physics of any kind has powerful appeal. Certain branches of the discipline possess some very straightforward explanations of our natural world, taming complex phenomena into predictive formulae, even postulating the existence of one grand physical truth that goes by the telling name The Unified Field Theorem.

Physics derives some of its attraction from the fact that human beings enjoy straightforward explanations. The reductionist tendencies of physics have produced some extremely impressive results, and for many centuries people have admired the accomplishments of the discipline. But that just shows the power of the forthright explanation. Even for the medieval soldier, obtaining the recipe for the ointment of war might have initially been popular because it was so much like branches of physics, linear and predictable. Do this, do that, and you get the results you want. A lot of biologists call such simplicity "physics envy".

Now just why do we biologists use the word, "envy"? – because biological scientists do not live in the cut-and-dried world of the physicist. And we'd *love* to, creating a professional jealousy, and also creating the term. The problem with elementary explanations applied to biological systems is that, to put it mildly, the explanations are fraught with variables.

Biological systems are not simple, even if you are looking at seemingly uncomplicated organisms like those uncooked earthworms. The armies that used *Unguentum Armarium* soon found that they did not get the results they wanted, and had to cook up some other crazy explanations. It was another six hundred years before a useful treatment for battle wounds became available.

A good lesson learned

Six hundred *years*? That's the delay between this recipe and the revelation that germs cause disease. Why did it take so long? The answer is that it took a long time for medical types to learn from our physicist colleagues. In order to make the same kinds of advances, those conducting biological research had to apply an intellectual rigor known almost solely to Isaac Newton and his friends. In the case of medicine, this meant starting over. Though a mortally wounded medieval warrior might not have been interested in how the metal, dirt and human flesh interacted, the great contribution of twentieth century science to medicine was to show that this interaction does matter. Biological systems were found to obey the laws of physics and chemistry. If you wanted to cure something, you would have to understand how those rules were played out in the systems under study. It's astonishing to see how recent this idea really is. It wasn't until the last fifty years of the twentieth century that you stood a better chance of getting better if you went to the doctor than if you stayed at home and just waited out what ailed you!

There is some understanding to be had in our six-hundred-year wait. When we began looking under the hood of biological organisms, we were instantly amazed and profoundly discouraged. Biological systems possess some of the most complicated, exquisitely designed mechanisms ever encountered by science. Just the source code – the genetic information inside human cells – has such a volume of detail that if you lined up all the DNA inside a single human being, you would create an unbroken queue stretching to the Sun and back one hundred and fifty times. There are around one hundred thousand genes inside just one of those cells, and we have isolated only a tiny percentage of them. In other words, there is a lot of stuff inside us which we don't know very much about. And that previous sentence describes the *present* condition, not the comparatively meager knowledge of past decades.

When you begin to talk about molecules and behavior, you pile such a level of complexity upon this already staggeringly difficult genetic reality that the task almost defies comprehension. That's because you are necessarily forced to work with the most convoluted and talented groups of cells in the known universe, the human brain. And by examining behavior, you are looking at one of the brain's most sophisticated functions. Consider just the numbers: there may be as many as a million million cells in the human brain, packed so tightly that a pea-sized chunk of tissue contains almost two miles of them. We call these cells neurons, and figuring out how the brain uses them would seem an impossible task even if they were just lined up end to end like snippets of wires, each associated by a single connection in a nice physics-kind of way. Unfortunately, the design specs are not that simple. A single neuron can have connections to literally thousands of other cells. Moreover, we are finding that these connections are exquisitely organized. Most form associations with immediately neighboring cells, creating local area networks of neurons (with infrequent associations with more distant areas farther away). The record is a cell type known as the Purkinje cell, located deep within the brain, possessing more than one hundred thousand different connections to various local groups.

To make matters worse, we are also finding that there are many different kinds of associations, some turning certain specific connections on, others turning things off. We are also learning that individual neurons can read these on/off patterns in aggregate ways, making intelligent decisions based on the shifting patterns being detected. This means that a single neuron acts less like a wire and more like a computer. If you can imagine one million million computers all connected in parallel, capable of talking to each other at around nine hundred meters or three thousand feet per second, you have an idea of the enormous firepower that lies just below our scalps. And when you understand that we have never isolated even one tiny active human neural circuit in a lab dish, let alone delineated how huge battalions of neurons work together, you have an idea of just how ignorant we are.

That's an important point, for even if we knew precisely how every gene worked in every cell of the brain, we still wouldn't know about the connections. Since in the brain connections are everything, we don't even know much about *obvious* behaviors, like fight-or-flight responses, let alone subtle ones, such as pair-bonding and divorce. The state of the research effort can be put this way: we have a working knowledge of a tiny fraction of human genes, and we have an even smaller working knowledge of the computing power embedded in the neurons that use these genes. No wonder biologists have physics envy!

So what does this have to do with behavior?

Given this paucity of knowledge, you may at this point be wondering what the rest of this book is going to be about. That's a fair question, but I bring up this idea of complexity to underscore a single point about human behavior, and to tell you why I get so mad at the headlines. Let me give an example, which takes the form a transcript of a radio interview I did awhile back. The subject was the impact of genetic engineering, involving a half-hour interview with call-ins. A woman's voice came on the line:

> WIFE: My husband said that its okay for him to see other women. He read there is a gene that makes men want to . . .
>
> RADIO HOST: To have their cake and eat it too?
>
> WIFE: No. He just said that it was natural for guys to go out with other women, 'cos there's something about spreading their genes, making babies, or whatever.
>
> HUSBAND (comes on the line!): That's not what I said! There's this article in the newspaper and all it said was that males have this natural urge to be with lots of women. It sort of said it was in our genes. Isn't that right, Dr Medina? Isn't there some kind of natural gene for this kind of thing?
>
> ME: No. There is no gene for promiscuity. Some people are trying to explain why women have menopause and men do not, but that's not a . . .
>
> WIFE: See! It's not your genes, its just because you don't have any . . .
>
> HUSBAND: Have any what?
>
> ME: Oh dear.
>
> RADIO HOST: I think we need to take another caller.

Do you see what I mean? This interview shows how easy it is for even well-meaning articles to be misinterpreted. When we first began to understand that gene activation could contribute deeply to human behavior, a lot of people began to go overboard with the nature side of the famous nature/nurture debates. Every emotional outburst, every violent behavior, every sexual peccadillo became explainable by our DNA, and the perception of behaviors and motivations began their slow slide into oversimplification. With this sad ignorance came all the inaccurate baggage of an *Unguentum Armarium*; and great predictions were made about our future ability to design our moods and understand why people eat chocolate, as well as the imminent extinction of mental illness. Because we knew a few things, we projected that we knew many things, and the dangerous assumption that everything could be explained biologically came into vogue. The enthusiasm completely ignores how little we know about the physical side of the products of our brains, and many have begun overinterpreting even the smallest stride in the field.

There is another reason why I wish to bring up this complexity. Since so much of the biology of human behavior is not understood, there is great controversy concerning exactly what it is. This controversy started with the people in the position to know best, the professionals who study behavior for a living. Here's one example.

All but one of The Seven Deadly Sins mentioned in this book have emotional content and, indeed, much of this book's pages try to explain the involvement of genes in emotion. But that's not an easy task from the start; there are many experts out there who completely disagree on the nature of emotion. There is even one professional who believes emotions don't exist. Consider the following quotes, the first from Dr Joseph LeDoux, author of the insightful book *The Emotional Brain*, and a recognized expert in the neurobiology of human emotion.

> I view emotions as biological functions of the nervous system. I believe that figuring out how emotions are represented in the brain can help us understand them. This approach contrasts sharply with the more typical one in which emotions are studied as psychological states, independent of the underlying brain mechanisms
>
> *(LeDoux, 1996)*

He then goes on to state what he thinks an emotion is.

> My idea about the nature of conscious emotional experiences, emotional feelings, is incredibly simple: it is that a subjective emotional experience, like the feeling of being afraid, results when we become consciously aware that an emotion system of the brain, like the defense system, is active.
>
> *(LeDoux, 1996)*

Implicit in these quotes is the idea that emotions exist in such a form that they can be studied scientifically; that there are actual brain mechanisms for emotions just waiting for the keen experimental insights of researchers to uncover. Nothing wrong with that, except that this enthusiasm is not universally shared. Consider this quote, taken from another academic professional, Paul E. Griffiths, author of an equally fine book *What Emotions Really Are*.

> (The general concept of emotion) needs to be replaced by at least two more specific concepts. This does not necessarily imply that the emotion concept will disappear from every day thought . . . Concepts like "spirituality" have no role in psychology but play an important role in other human social activities. But as far as understanding ourselves is concerned, the concept of emotion, like the concept of spirituality, can only be a hindrance.
>
> *(Griffiths, 1997)*

Can you sense the conceptual discrepancy between these two important individuals? Is there enough evidence that emotions exist for them to be studied by a neurologist? Alternatively, are they just an out-dated idea so badly in need of a conceptual tune-up that the notion should be erased before we go trotting off to the lab bench? Further reading of both men's ideas shows that they agree on many important points. However, they disagree on so many other issues that one is left either scratching one's head or nodding in agreement with the fact that they disagree for the precise reason that we must be cautious: we know so little about any part of brains and behaviors that we are still asking basic, what-is-it-really, type questions. Without a responsible framework, there's no way to ascribe individual genes to individual behaviors and hope to relay an important truth. And that, precisely, is the point I am trying to make.

There is another reason to be cautious. There is an area of the brain called the cortex, a skin-like covering of our brains where most of our sophisticated "human" functions are processed. The cortex helps us in our ability to generate and process language, and allows us to think about how to manipulate our environment using our useful opposable thumb. It also gave us a remarkable characteristic: it enabled us to create a social organization that could manipulate its environment. This organization was so unique to individual people groups that the concept of culture was born. The ability to use our hands with our talented brains gave us physical art, types of houses, specific religions, and discrete behavioral protocols. These protocols became so powerful that their mores and customs could reconfigure the very organ that started them (we will talk more about this phenomenon, often called neural plasticity, in the last chapter). In other words, culture plays an enormous role in the shaping, even in the origins, of our behaviors. Moreover, except in the broadest sense, there is nothing *genetic* about it.

A complete explanation of emotion and behavior cannot omit this enormously important social input. Since this book focuses on biology, we will not be addressing this critical role for culture, and hence our discussion is imperfect. I mention it here, however, simply because it adds another layer of complexity to an already convoluted task. Also, we need to understand that the subject *is* this complicated as we attempt to address phenomena such as sexual arousal and anger, for example. In fact, most biologists shy away from the social inputs, not because we believe they are unimportant, but simply because the task is too overwhelming to try to integrate the inputs successfully in a test tube. I will remind us of the importance of social forces in shaping behavior in the next chapter.

So what are we going to do?

It probably sounds fairly radical to say that there is no such thing as emotions, or that behaviors are so plastic that social forces can actually alter their contours, if not change them outright. We are so familiar with our own behaviors, and are so depend upon their presence, that life would seem impossible to live without them. And because emotions are so deeply rooted in us, our tendency is to quickly ascribe their origins to genetic roots. We even get these feelings at an early age, a fact brought home to me one fine morning when my son was about four months old. To understand the following story, you must know that my son was born with red hair, and lots of it (he actually had to endure three haircuts before his first birthday).

One rainy morning in Joshua's first autumn, my son began to discover this tangled mop of protein squatting on his forehead. He seemed to be asleep, and at first he just touched his hair softly, patting it drowsily here and there when he could get his arms under control (no mean feat for this four month old). But on this rainy fall morning, I watched him touch his hair as usual, make a fist, and then pull at it with all his might. This of course hurt, and at first he looked wild-eyed, as if he were afraid, and then he howled in pain, clenching both fists, crying at the top of his lungs. This clenching included the hand holding the hair, strengthening his grip, causing him to scream even louder, which caused him to clench even tighter and . . . you get the picture. Poor Joshua was caught in a vicious loop of emotional reactions, and I distinctly remember when he realized it – the quality of his cry changing from pain to what can only be described as anger. It took many minutes for Daddy to gently pry the source of his consternation away from his head. And when he let go, I saw him smile, relax his little hands and fall back to sleep on his pillow.

How intriguing to watch fear and pain and wrath and relief and happiness come into focus in my son's brain, like some biological film slowly being developed. Psychologists – as well as zoologists – tell us that this emotional development (or *whatever* development we should call it) is very important for a baby's overall growth, and has powerful reasons for coming into existence. A human's ability to feel fear, for example, is a critically important survival skill. Joshua will never develop the teeth of a jaguar, the physical strength of a grizzly bear, or the sharp talons of an eagle. These unfortunate facts mean one thing: if Joshua were forced to survive in the wild, he would have to become something of a coward, doing lots of avoiding, learning to become afraid of life-threatening situations, becoming angry when his environment is disorganized, happiness when life settles down, all in the hope that he can make it through the next day. In this view, the ability of the brain to conjure up specific motivations is a talent meant to keep it alive, pure and simple.

The previous sentence makes a powerful judgment. It says that one of the chief job descriptions of the brain is to act as a survival organ, and that the organ's ability to motivate is nothing more than another powerful tool designed to help us pass on genes. The fact that we are capable of complex behaviors simply means that we have chosen to grow fangs not in our mouths, but in our cortexes, that our physical strength is developed not in our muscles but in our neurons, and that it is our IQs, rather than our fingernails, that give us the intellectual talons sharp enough to claw out a place in this world. This evolutionary view, which I believe to be the proper one, reveals to us some real hope for understanding behaviors. We obtain a clue as how to write about the genes of motivation, even if we don't really know what they are.

How we will look at emotions and behaviors

One reason why this evolutionary context is an important perspective is that it allows us to make several predictions. Many researchers believe that our emotions and behaviors evolved to give us a fighting chance against the saber-toothed tigers. If our brains really did develop specific types of motivations to survive in a hostile world, then the neural architecture mediating those behaviors must have been crafted in the harsh crucible of natural selection. Since natural selection is a physical phenomenon, these neural-based behaviors are necessarily bequeathed with certain objective components. And because science is the discipline that can measure objective components, we can safely use science's powerful tools to understand certain things about behaviors. In other words, studying The Seven Deadly Sins is not a waste of time in a science-oriented culture, for the effort will make eventual sense in a test tube.

This evolutionary context idea reveals not only the tools we may use, but also an important reference point. It is obvious to anyone who has ever had a household pet that animals also possess behavioral repertoires, some deceptively familiar to our own. From fear to anger to lust to hunger, other creatures with complex brains possess motivations that cause them to act in certain ways. Of course we don't know if they are the same feelings that we experience (without a cortex as sophisticated as our own, I do not even know what relevance the word *same* has here). However, their presence is unmistakable, and, as with any physical phenomenon, worth questioning.

We know, for example, that animals have to satisfy certain environmental conditions in order to survive, the same conditions we humans have to satisfy as well. Many have sophisticated brains and they all have genes, some looking very similar to our own, and, in many cases, functioning like ours do. It is even possible that the brain systems that generate vital emotional behaviors are conserved throughout the animal kingdom. If that's the case, then studying the link between genes and behaviors in animals is not a futile effort. Rather, animal studies can work like flashlights, illuminating certain tissues in the vast thicket of human cells, selecting the ones researchers should study first. As we shall see, using animals as an initial referent has met with a measure of success, and genes involved in specific behaviors have been isolated. Because there are also a fair number of pitfalls in this approach, we will return to this idea of humans and animals and behaviors again and again throughout this book.

A few postulates

But what about the complexities we mentioned previously, as well as the basic questions? Can we leave conceptual conflicts such as those between LeDoux and Griffiths unresolved as we hurtle headlong into The Genetic Inferno? Given the convoluted nature of behavior, we will need *some* kind of operating framework upon which we can build, if our discussion is going to make any sense at all.

It helps to start with a few postulates. I will state flatly that I believe specific motivations in humans exist as biological entities in our brains, whether we choose to call them emotions or not. And I have chosen to use LeDoux's bipartite model of conscious emotional experiences, such as feelings, for our framework to explore the biology of most of The Seven Deadly Sins. This isn't because his model explains everything; indeed, both LeDoux and Griffiths would readily admit that there is probably no one monolithic archetype for all human emotions (or mental states, or whatever one chooses to call such behaviors). Before we move to our discussion on biology and Dante, and with apologies to Dr LeDoux for any oversimplifications, I would like to explain the components of the model we will be using.

As you recall, LeDoux believes that emotional experiences occur when an "emotion system" of the brain is stimulated, and we become consciously aware of its activation. In his view, there are two parts to the experience. The first, the emotion systems, are those discrete collections of tissues, neurons, and genes involved in a specific response, such as fear, hunger, or sexual arousal. These systems give the emotion its particular identifier, Dante, a category for his particular sin, and us a running subject for each chapter.

That's the comparatively "easy" part of the model. The next step is not so straightforward. These emotion systems are connected to an idea so complex that nobody really knows what it is, except, of course, that it exists. This idea is the concept of conscious awareness, and is the second component in LeDoux's model. It is frustrating that so little is known about the neurological basis of consciousness, because, in many ways, LeDoux's framework is really an outline of how conscious experiences occur in people. We are not totally in the dark, fortunately. Researchers know that certain biological components have to exist for consciousness to occur, even if we don't know exactly how those components work together. It is like seeing an incomplete list of ingredients for a recipe without cooking the food; we might be able to imagine its flavor, but we have no way of understanding the taste of the finished product.

It is my intention in the coming pages to describe how this model might function in each Deadly Sin. We will spend the majority of our time talking about the biology of a particular brain system, usually starting with large tissues and then moving to tiny genes. We will use our discussion of minds and brains at the end of each chapter to talk about a specific ingredient of the second component of emotion, the notion of consciousness.

Though the various components are unfolded in a serial fashion, the listing is by no means a complete explanation. Many researchers, including myself, believe that at this point in the technology, *nobody* can give a complete discussion (we wouldn't have space even if it were possible, for vast numbers of books have been written on just this subject, most coming to the same the-jury-is-still-out conclusion). It is left to the reader to imagine just how the various ingredients might coalesce to give us the experience of awareness.

Taken together then, these caveats represent a few of the issues we must remember if we are to deal responsibly with the subject of behaviors and genes. I do not mean to be discouraging here. I cherish simple answers as much as anyone does. The problem is that we were not made simply; the human being is one of the most complex works of engineering that exist, and our specialized cortex which assists those behaviors is one of this complexity's most intricate tasks. We are beginning to understand a few things, making some tentative associations based on what we know. But we have in no way discovered the DNA responsible for some people's enthusiasm for eating chocolate. We don't even know how appetite is generated, and, if you take the view of some researchers, whether appetite even exists as a measurable phenomenon.

The biological background

With these caveats in mind, we are ready to tackle some important biological background. Then we will give a brief history of *The Divine Comedy*, and head straight off to The Genetic Inferno. However, there are three groups of information we need to review briefly before we enter the strange world of human behavior. These include some facts about cells, and some facts about neurons. Once again, please remember that this book was not written for professional scientists, but for people with all the biological background of literature and language majors. We will take our time to explain things, use plenty of metaphors along the way, and clarify the important points with illustrations.

The cell

The most important thing you can say about cells is that human beings have lots of them, sixty to eighty million million, to be inexact. They come in all shapes and sizes, and perform a myriad list of functions (see Figure 1). Some cells are round and look like beach balls. Some cells possess electrical activity and functions such as stimulating muscles or causing brains to write famous fourteenth century books. These cells are called neurons, as we mentioned, which are cells that look something like a scared mop.

Whether beach ball or mop, all cells possess a layer of fat around their outer surface. Essentially a greasy rind, we call this layer the outer membrane. Inside the membrane is an area filled with salt water and known as the cytosol. This saline-like environment is criss-crossed with all kinds of structural supports, forming a skeletal framework termed, unsurprisingly, the cytoskeleton. The configuration of this cytoskeleton, much like our own larger model, helps decide the three-dimensional shape of a given cell.

A skeleton is not the only thing a cytosol contains. Located inside this salt-water environment is a spherical-shaped structure known as the nucleus. This is the command and control center of a cell, and it gets this lofty title mostly because of a single biochemical. Packed into the nucleus' tiny volume is about 1.8 meters of the most famous molecule on earth, DNA (short for deoxyribonucleic acid). This large amount of DNA makes the nucleus a very crowded place, equivalent to putting thirty miles of fishing line into a space the size of a cherry pit. This packed DNA does not form a single contiguous line, however. Human DNA comes in forty-six different segments, just as an encyclopedia might be divided into forty-six volumes. Each of these volumes is called a chromosome. The DNA is further subdivided into particular regions that we used to call "traits" but are more commonly known as genes. As we shall see, genes encode information for molecules that perform an almost bewildering number of functions. Some of these molecules play an important role in human behavior.

The most fascinating thing about the sixty to eighty million million cells that make up your body has to do with the word "redundancy". To see what I mean, consider the following true story.

You can actually remove certain tumors from the inside of people and find fully formed fingers deep inside them. You did not read that wrong; real live differentiated tissue exists in these nightmarish cancers, and fingers are not the only things there. You can find teeth, and skin that has hair on it, and bones and cells that beat like mini-hearts and eye tissue . . . in fact, the entire range of human tissues can be found in these tumors. These cancers are called teratoma tumors, and are a source of unparalleled delight to many research scientists.

A typical animal cell

There really isn't any such thing as a typical animal cell. Shown on the left is what many neural cells look like. Shown to its right is a generalized human cell.

A CUT-AWAY VIEW

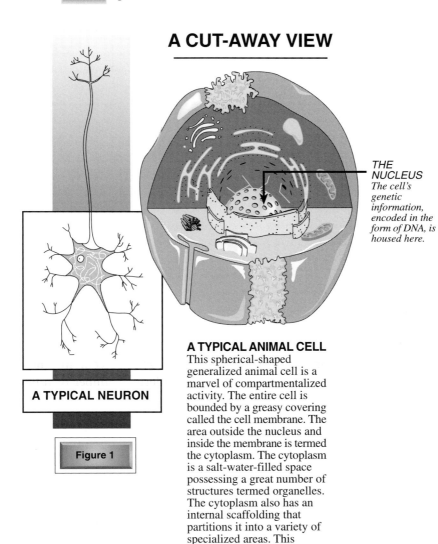

THE NUCLEUS
The cell's genetic information, encoded in the form of DNA, is housed here.

A TYPICAL NEURON

Figure 1

A TYPICAL ANIMAL CELL
This spherical-shaped generalized animal cell is a marvel of compartmentalized activity. The entire cell is bounded by a greasy covering called the cell membrane. The area outside the nucleus and inside the membrane is termed the cytoplasm. The cytoplasm is a salt-water-filled space possessing a great number of structures termed organelles. The cytoplasm also has an internal scaffolding that partitions it into a variety of specialized areas. This scaffolding is termed the cytoskeleton.

That's correct. *Delight.*

From a distance, it appears that the cells are trying to become a human being, but have lost some critical signal and instead collapsed into what looks like a robbed grave. They aren't trying to become anything of course, but are simply responding to an extraordinary occurrence, a fact that has to do with the word redundancy mentioned several paragraphs ago. It turns out that every cell in your body, including those that could eventually become a teratoma tumor, possess all 1.8 meters of human DNA. That means they also possess all one hundred thousand genes, the entire blueprint for the manufacture of one human being. That means that the cells in your cheek have the genetic instructions for your heart, your liver, your big toe, in fact every tissue in the body. Sixty million million copies of everything, truly an exercise in redundancy.

If that extraordinary fact is true – and it is – you can ask an important question of your mouth: why is a cheek cell always a cheek cell? If that cell truly has all the genetic information to make every tissue, why isn't every tissue in your cheek? Even if you wound the inside of your mouth, you won't grow back a foot, but rather other cheek cells. So not only is there selectivity, there is also memory. How does it all occur?

The answer to that question is beginning to be understood at a refined level, and it is the reason why scientists are so delighted. It turns out that all the genes necessary to make a cheek cell are turned on in a cheek cell, and all the other genes are repressed, rendered nonfunctional. The same is true of a liver cell, where all the genes necessary to make a liver function correctly are active, and everything else (including any cheek cell genes) is turned off. This idea of turning genes on and off is exciting because we are learning how nature does it, and are in kind learning to turn them on and off ourselves. In many cases, the teratoma has shown the way, for it has learned how to turn on genes in such a selective order that entire structures are formed. And if the tumor can do it, so can we.

The idea of selective activation of genes inside cells is very important for our understanding of behavior too. As we'll encounter shortly, neurons "learn" things by turning certain genes on and other genes off. Different cells respond to specific hormonal cues, from sex to fear to hunger, by turning genes on and off. It is important that we understand how richly selective gene activation is before we talk about their roles in specific behaviors.

A lot of nerve

With the architecture of cells and a little bit about activation under our belts, we are ready to talk about our last biological subject. I want to shift our focus to a single cell type, those scared-mop-like cells of neurons, and talk about how they work. Any understanding of behavior at the molecular level begins with the fact of neural activation.

It may seem simplistic to say so, but our entire bodies run on electricity. This is because our neurons run on electricity, and they control almost everything that makes us uniquely human (as well as almost everything we share in common with lower animals). To study how neurons become activated, we have to know a little something about how they interact with this electricity.

As you remember from high school physics, the heart of electrical phenomena consists of the presence of charges. Like the yin and yang principle of traditional Chinese thought, there are two types of charges, positive and negative, at once fully opposite and fully complementary. You might also remember from high school that atoms can carry either positive or negative charges. If you can keep atoms carrying opposite charges separate from each other, you create what is termed a polarized condition (positive being on one side – one pole – and negative being on the other). If entities carrying one charge are allowed to invade the area occupied by the other charge, the situation will lose its polarity. Not surprisingly, in this situation, the venue is said to be depolarized.

This idea of separating charges and then allowing one type to invade another lies at the heart of neural signaling. It turns out that neurons are sitting in a saltwater bath of charges, and they have developed mechanisms to keep some charges outside their membranes and other charges inside. One positive atom that neurons absolutely love to keep inside their cytosols is potassium, an atom found in abundance in bananas. It turns out that we have lots of atoms of potassium too, used not only by our nerves but by many cell types.

Our neurons are possessive over the atom only when they are at rest, a situation that makes the cell *polarized* with respective to potassium (lots on the inside, very little on the outside). If a neuron is going be stimulated, something extraordinary happens. The cell actually lets its positive potassium atoms deliberately escape from the cytosol, among other things. Neurons do this by opening up tiny molecular faucets embedded within their greasy membranes, and letting the potassium dribble out (we don't really call these membrane-bound molecules faucets, but ion channels) (see

Figure 2). As the potassium escapes, the inside of the cytosol becomes less positive with respect to potassium, just as the outside becomes more positive. The neuron is said to *depolarize* as a result. If enough potassium gets out, the neuron is electrically stimulated and is said to have "fired". As you might suspect, a great deal of energy has gone into figuring out exactly how those faucets are instructed to open.

The firing occurs all the FROG time. It is happening right now as you read this sentence. Did you just see the word frog? Of course you did. Close your eyes and you can still see that I stuck this odd word in the middle of the first sentence. Your ability to remember this seeming typo depends upon specific neurons firing, letting go of their potassium and storing the data. Such depolarizations occur with every learning experience, and indeed with every behavior humans possess. That's what we describe here initially; we will revisit the subject, adding to our knowledge, as we go through the book.

We are almost done with our background introductions. Before we get to The Seven Deadlies, however, we need to mention some information on the organizing principle of this book, comprised of selections taken from *The Divine Comedy*. This will give us our opening into the specific subjects of human behavior. Let us therefore remove ourselves from the unfamiliar and obscure world of the microscopic molecule and describe a world equally unfamiliar and obscure, though not microscopic at all. We are going to visit one of the great cities of fourteenth century Europe, Florence, and say hello to one of its most famous writers, Dante Alighieri, who we find currently languishing in exile.

A few words about Dante, *The Divine Comedy*, and human behavior

It must have been the ultimate indignity for a man of Dante's stature to be kicked out of his hometown. The year was 1302, and Dante Alighieri, already an established poet, held high public office in his beloved Florence. Unfortunately, he belonged to a party on the losing end of political power. His rivals eventually confiscated Dante's possessions, sentenced the poet to permanent exile, and to capital punishment if he should ever come back to Florence. It was in this exile that Dante wrote his masterpiece *The Divine Comedy*, finishing it just before his death in 1321. His bitter view of the civic forces that banished him runs through the book like the Arno. So does a great deal of medieval thinking about human behavior. *The Divine Comedy*

The electrical nature of nerves

Neurons live in a world of electrical charges. Keeping charges inside or outside their cytoplasm determines neural activity. Here's how a typical neuron works.

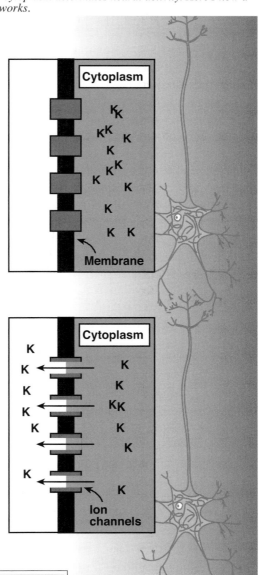

IN THE RESTING STATE
Neurons that are not actively involved in communicating information are said to be "at rest". This means that some charges are kept outside the cell and some are kept inside the cell. Shown on the right is the example of potassium (represented by the letter "K"), a positively charged atom. Neurons in the resting state keep potassium inside their cytoplasm. The neuron is said to be "polarized" with respect to the atom.

IN THE ACTIVE STATE
When a neuron is electrically stimulated, it becomes "active". In this example, proteins called ion channels open up and allow potassium ions to escape into the outer environment. Because positive charges are now moving from one place to another, the situation quickly becomes "depolarized" with respect to potassium. When certain other changes occur, the depolarized neuron is said to have fired.

Figure 2

is a book about the afterlife, and how the actions of the living have spiritual consequences after death. The book is divided into three overall sections and is called a comedy, not because it is funny but because it has a happy ending. The three sections are written like a travel editor's journal, describing Dante's personal tour through hell (*The Inferno*) through Purgatory (*The Purgatorio*) and finally through Heaven (*The Paradisio*). Though each as long as a book, these three sections each comprise giant poems and are divided into various Cantos, the principal divisions of a long poem. Dante writes these Cantos in *terza rima*, a three-line stanza style invented by him. When we quote from *The Divine Comedy*, we will be using the translation of John Ciardi (cited in the references to this Introduction), who attempts to faithfully reproduce this *terza rima* in form in English.

In the opening pages of *The Divine Comedy*, we find out that Dante will not be alone on his tour. Rather, he is guided through most of the Christian netherworld by a former pagan (!), the ghost of the Roman poet Virgil. The travelers also meet various members of the afterlife trapped at specific levels, who often describe their behaviors, the reasons for their entrapment, and the consequences they endure as a result. Of course this is a tale of allegory and metaphor, and the brilliant layering of meaning embedded in Dante's text has piqued the curiosity of medieval scholars for years and years (many believe that Dante actually wrote *The Divine Comedy* as a reaction to his own life's circumstances, as an attempt to explain the problem of pain in the presence of a loving God). Inadvertently perhaps, a great deal about how a medieval mind views human behavior is revealed in the pages of *The Divine Comedy*, and, I have to tell you, it is remarkably fun reading.

The Seven Deadly Sins are described in formal fashion in the middle section of the book, *The Purgatorio*. Dante envisions Purgatory as an almost indescribably huge mountain, erupting from an infinitely vast sea, and looking something like a wedding cake. The reason for this layering is that the mountain is divided into a number of terraces or cornices (an architectural term for a molded projection, usually crowning a building or wall). The top seven levels are reserved for those souls guilty of one of The Seven Deadly Sins, and it is the task of Dante to walk along each cornice, describing what he sees. The top of the cake is the Seventh Cornice, holding the inmates bothered in life by the problem of sexual lust. The Sixth is for those who struggled with gluttony, the Fifth for the greedy and the Fourth for the slothful. The next three cornices are comprised of people guilty of wrath, envy and pride respectively.

As you might suspect, we are going to use the adventures written in *The Purgatorio* section in an attempt to repel our way down the great mountain of genetic and behavioral research. This research will be described in a top-down approach, as previously noted, starting with large groups of cells and ending with small groups of genes. We will approach our journey through Purgatory in the same style, starting at the top of this wedding cake – with lust – and working our way downwards, until we get to pride. (One professor whose lectures I attended for background information described *The Purgatorio* this way too, and so our imaginary professor at the start of each chapter will follow suit). It must be noted, however, that the Poets journeyed through Purgatory in the text in exactly the opposite way. Dante and Virgil actually started at hell, traversed Purgatory by going from the First to the Seventh Cornice, and then into heaven.

With these biochemical, literary and organizational comments in mind, we now have enough background information to start our tour. And that brings me back to the subject of the beginning of Introduction. I think about my little boy, now fourteen months old at this writing, hearing his sandbox-generated laughter as I pound out these sentences on my word processor at home. I wonder about the genes his mother and I gave him, and the distance between his synapses and that incandescent smile that so completely lights up his face (and ours). He's throwing clods of dirt on the lawn at this moment, and I laugh too, already observing a familiarly mischievous gleam in his eyes. I stop myself before I hear the sound of water filling up a station wagon, and my mother's voice telling me about my dad . . .

All of these wonderful things give us a compelling reason to peer into the chasm of human behavior. It's an odd place, where survival sharpens the brain into a tool that can create swords and become afraid of them at the same time. Where the under-the-skull similarities between animals and humans immediately say a great deal about our natures and absolutely nothing at the same time. And, most of all, by turning thoughts into electricity, it is a place where the simplistic notions of human behavior burn to a crisp in the heat of The Genetic Inferno.

CHAPTER TWO

Lust

luxuria

"So, one before the other, we moved there
Along the edge, and my Sweet Guide kept saying
'Walk only where you see me walk. Take care.'"

-Canto XXVI, The Purgatorio

"Slap on some suntan lotion and put on your lightest clothing, everyone", the old professor cleared his throat, looking over his glasses at about one hundred undergraduates, "You are going straight to hell." The class burst out laughing. The professor smiled, and then seemed to notice a young man with red hair looking impishly – but persistently – at another member of the class. The old man stared right at him, "We will start this journey backwards," he declared, "and like a lot of people's slow ride to oblivion, we will begin with your genitals. The seventh level of Purgatory is reserved for those enslaved by the great sin of sexual lust!" The red-haired student blushed, and the class laughed again. The professor broke out his copy of *The Purgatorio*, the middle section of Dante's famous *The Divine Comedy*, and started lecturing at the speed of light . . .

The good mentor was not angling for a sexual harassment suit, of course, but was merely giving us a quick introduction to the odd world of Purgatory Mountain, the top cornice in Dante's description of middle after-life. The professor started his lecture by describing who was accompanying Dante on his tour of the Seventh Cornice. There was his ghostly guide, the Roman poet Virgil, companion to Dante through much of his journey. There was also another, more temporary colleague, a long-winded ghost named Statius, actually a resident from a level of Purgatory several cornices below.

To reach the Cornice of Lust, the poets had to climb up a narrow staircase. "It was a long, narrow climb," the professor explained, "and Statius takes the opportunity as they walk to talk about sex. He focuses mostly on the mechanism of human conception and the nature of the soul, a bit tedious for my money, especially considering the subject. Gives some interesting medieval points about human origin, though, with such great insights as the ability of blood to change into sperm, and the human soul coming only from males. Statius stops his monologue only when the staircase opens up to a ledge."

The ledge led to the cornice, of course, a narrow trail that ringed the top of Purgatory. The professor explained that the poets beheld quite a sight. On the left side of the ledge, clinging to the mountain like death, was a sinister rolling fog. This fog was an odd color, because contained within it was a roaring furnace, shooting walls of famous medieval flame up the side of the mountain. Somewhere a choir could be heard singing. Above the din, one could also hear the roaring waves of the ocean crashing against the mountain far below.

"Dante is scared, of course, especially of the fire," the professor continued, "but he summons his courage, looks directly into the flames and immediately discovers the source of the music. There are spirits being roasted in this furnace, singing and writhing and marching somehow within these flames. Their incantations are hymns with a mix of subjects. Sometimes they're about God's clemency, sometimes the sexual abstinence of biblical figures such as Mary. There's even mention of the Roman goddess Diana, and some kind of poisoned blood."

Though hideously torturous and uncomfortable to behold, the Seventh Cornice was not a lonely place for Dante. The professor went on to say that Dante recognized a few of the residents as thirteenth century poets, some quite famous in his day, and even struck up a dialogue with some of them. Virgil admonished everybody to keep "a tight rein" on their eyes, lest they too fall off the ledge and into sin.

What this chapter is about

Everyone makes it out of the cornice intact, though the professor explained that the Poets could not leave without themselves walking through a flame. Mentioned both as a purifying device and as an illustration of burning desire, the use of fire as a metaphor was as obvious to medieval readers as it is to today's scholars. In this chapter we are going to focus on the biology of the latter use. Obviously, we are going to talk about lust, focusing specifically on the biology of sexual feelings. We will begin with a few brain cells, move to a few hormones, and end with a few genes. Then we will discuss how this biology informs us about the emotions of sexual arousal, the real lust the Seventh Cornice describes. This is rather difficult to do, for it means addressing in part the distance between our minds and our brains, as well as notions such as consciousness. We are defining emotions, you remember, as an emotional system coming into conscious awareness.

Though there is a fair amount known about the substance in which these feelings are deeply rooted, we shall soon see that not much is understood about subjective lustful feelings. By the end, we will understand just how complicated such systems are, and that our attempts to organize and label what we feel do not always jibe with biological reality. In fact, a surprise awaits us: many of our reactions that make up the feelings of lust may lie outside our ability to perceive them. In fact, there may be no such thing as the subjective *emotion* of lust at all.

With that cryptic sentence, let us start our discussion of lust with two critical questions. Exactly what is the biology behind sexual feelings? How does it create subjective feelings of desire? As will become obvious, finding the answer to the first question is comparatively straightforward, and finding the answer to the second practically impossible.

The danger of simplification

It is easy to discriminate intuitively between the appetitive and consummatory aspects of human sexual responses (even though it's somewhat difficult from a scientific point of view to mark their boundaries). We are about to focus on only one portion of the human reproductive experience – the appetitive aspect – without trying to distinguish it from other definitions of sexual feeling (like libido, for example, whatever *that* is). This appetitive aspect we will simply call sexual arousal.

At one level, the experience of sexual arousal seems simple enough. A stimulus occurs, a pleasurable feeling appears, and we go away with an ache in our hormones, a grimace or a smile. The origins of those feelings appear equally simple and straightforward. If someone touches us in a highly erotic way, we may become stimulated sexually, perceiving the emotion of sexual arousal even if we weren't previously thinking sexual thoughts. Conversely, we might generate highly arousing sexual feelings in the privacy of our thoughts, and our bodies may then respond to the dictates of this internally supplied stimulus. Arousal appears to be a simple two-laned highway, one lane arising from the body and going to the brain, the other lane arising from the brain and going to the body.

Such goings on can feel so monolithically uncomplicated that some people simplify the concept when they inquire about its origins. They may ask, "what region of the brain possesses this highway?" or, even more simply, "what is the gene responsible for these feelings?" As innocent as these questions may sound, they make surprisingly little sense from a biological point of view. To understand what I mean, consider the following illustration.

When Dante was twenty-six years old (1291 AD), the construction of a beautiful minster began far away from Florence, in the English town of York. An elegant stained-glass window commissioned for the building was completed in 1338. It came to be known as The Great West Window, and still survives in its lofty perch today. I have had the distinct privilege of visiting the minster and gazing up at The Great West Window. It is a huge work of art – over one hundred and eighty-five square meters or two thousand square feet – among the largest examples of medieval stained glass left in the world. When the sun

shines through it, the pigments create an almost unbelievable aerial kaleidoscope, filled with fantastical shape, fine tracery and every color in the
rainbow.

The reason I bring up the subject of complex stained glass in a chapter
on sexual biology is as follows. A child who had never seen The Great West
Window, or even medieval stained glass, could ask a naive and simple
question, " what color is The Great West Window?", perhaps imagining that
one true pigment existed to visually organize the object. You have no
difficulty in detecting the awkwardness of this question, of course. No *one*
color is responsible for The Great West Window. Rather it is a pageantry
of many shapes, traceries and pigments. We experience a singular medieval
composition only as an emergent property of its constituent parts, working in a coordinated fashion to fill one area of a church.

The same division of content in stained glass is also true of the biology
of sexual arousal. There is no one true gene involved in sexual arousal –
even if it is perceived as a monolith of pleasurable feeling – just as there
is no one true color in The Great West Window. Nor is there only one system involved in creating the sexual drives we experience. The feelings of
sexual desire are best understood as an emergent property of at least four
interlocking physiological systems, at least eleven different regions of the
brain, more than thirty distinct biochemical mechanisms and literally
hundreds of specific genes supporting these various processes. There is a
system of tissues involved in the creation of lust, but no one true region or
one true gene is responsible for the response, not to mention the subjective
emotion that derives from it.

Something we won't talk about

Though we will not address it at great length, human sexual processes cannot be divorced from their social contexts any more than the meaning of
the images in The Great West Window can be divorced from the century
in which they were painted. Those four systems, eleven brain parts and
thirty biochemical mechanisms have been forced to spread their sexual
fires within a given culture, sometimes literally for centuries. The end result
is that different people groups in time and space end up experiencing their
arousals in very different ways. These cultural forces influence the very
framework – if not the nature – of the process itself. There are thus many
reasons why we cannot ask "what is the single brain part or single gene
responsible for sexual hunger?" and ignore the influences of culture. There

are simply too many tissues – and issues – involved. There is more on social forces at the close of this chapter.

So what do we do?

Considering the nature of the task, do we then just throw up our hands and refuse to explore any biological input into human sexuality? Notwithstanding the prior caveats, there's plenty we can do. Without over-simplifying the biology or subordinating the impact of culture, this can be said: we survived in this hostile world because we were enthusiastically willing to hurl our genes clear into the next generation. Such enthusiasms have solid evolutionary themes coursing through them, themes amenable to biological inquiry and scientific explanation. We will start with descriptions of two neural systems, move to what are termed genitourinary and endocrine systems, and end with the role of circulating hormones.

The matter of nerves

This is the most obvious thing I will say in this chapter: at the most basic level, sexual feelings involve the use of human nerves. If someone is going to experience a sexual feeling, he or she is going to need a functioning brain, a somewhat intact spinal column and neurons connecting these structures to each other and to various organs that sense the outside world. We in the scientific community call the nerves that perform all these tasks the central nervous system. There is no doubt about the primary role of the brain in sexuality – it is easily the chief sex organ humans possess. Other neural structures serve as messengers to inform this lusty majesty about any sexual inputs currently being experienced.

So, just how is the brain wired to receive such interesting input? There are three ways. Streaming out of the brain and into the bodies of both males and females are lengths of nerve fibers we call cranial nerves. These connect the brain directly to sensory organs, those tissues mediating such vital sexual functions as vision, sound, breathing, heart rate, facial sensation, speaking, hearing, even tasting. Secondly, other noncranial neurons sprout from the spinal column, looking something like roots from a plant. These connect the chief sexual organ to other parts of body, informing our brains about temperature, body position and one of the most important sexual inputs – touch. The spinal cord also connects the brain with a third

set of nerve cells, the so-called motor neurons. These allow the brain to put
into action what it is feeling, working the various muscles in the body,
receiving input from them and issuing new commands.

There are other systems of neurons involved in sexual responses besides
the central nervous system. One system deeply involved in sexual feeling
is the so-called autonomic nervous system. Defined simply, the autonomic
nervous system connects impulses from the central nervous system to
various other tissues, and even glands, in the body. These cellular cables
regulate tissues (especially muscles) not normally under conscious control.
Such involuntary muscles include those involved in digestion, elimination,
circulation, and the secretion of specific chemicals from distinct glands
found in the body (more on that gland stuff a little later). This system is
further subdivided into the sympathetic and parasympathetic systems, as
further explained in the next chapter.

With all this emphasis on the brain, one might be tempted to think that
every sexual input must be routed through the brain in order for humans
to reproduce. Oddly, there are a number of sexual mechanisms that at first
appear to ignore the brain altogether. Several neural circuits function right
near our genitals, and a description of arousal would not be complete without a brief mention of how they work. These genital-specific circuits are
called reflex arcs, and they mediate the kind of actions experienced when
a physician taps our knees with one of those little silver hammers, the difference being that this input is strictly sexual. When our genitals sense
pressure, impulses are generated in specific sensory neurons that carry the
happy sensation away from our sex organs to a reflex clearing-house located in the base of our spinal column. The clearing-house then sends signals
back to the genital area without ever routing to the brain. These messages
communicate with specific muscles that control the walls of small blood
vessels in our sexually responsive body parts (areas of the penis in males
and areas of the clitoris in females, as well as surrounding tissues). It is
completely automatic, mediated by the system we just described and sounding a lot like the word, the autonomic system. Messages that travel via one
type of autonomic neural telephone line cause tiny muscles in our arter-
ies to relax, allowing increased blood flow. That's why blood can engorge
the penis and clitoris automatically during arousal. A comparable reflex
arc also controls blood flow to human nipples in both males and females.

A description of some of these autonomic responses and reflexes illustrates an important principle: there are aspects of sexual responsiveness
that lie outside our conscious control. For us to consciously experience
such impulses, they must be brought into our awareness, i.e., to our brains,

and from that awareness we may experience the subjective feelings of arousal. However, since these reflex arcs occur independently of even the most rudimentary comment from the brain, we can say an astonishing thing. Sexual feeling (perhaps even the emotion itself) is something that *happens* to us. There is a certain amount of processing of our responses that occurs outside our conscious control. Thus, while the brain is still easily the sex organ par excellence, we are not always aware what regions are trafficking in sexual information. Certainly the brain can generate its own sense of arousal, but the system it uses has components that are not subject to its orders. There is more about this automaticity at the end of the chapter.

These nervous systems are not the only set of tissues being alerted during sexual stimulation, of course. We now turn our attention to the next collection of cells involved in human arousal, the more familiar tissues of the genitourinary tract.

The genitourinary tract

The genitourinary tract, as you might expect from the name, comprises our genitals and the systems we use to eliminate urine. In males, the primary input organs are the structures surrounding and including the penis, and in females, structures surrounding and including the clitoris. We will talk about each of these organs and briefly discuss their role in sexual arousal.

The penis is a complex structure loaded with sensory nerve endings (the tip of the penis, called the glans, is especially sensitive to tactile stimulation). When one trunk of the autonomic system sends arousal messages to the organ, the blood flow entering the penis exceeds the blood flow leaving it and an erection is achieved. The stiffness occurs because of the particular architecture of the penis. The organ is composed of three spongy, cylinder-like structures that run lengthwise inside its shaft. When these fill with blood, the penis begins to rise like a hydraulic lift. Neurons within the penis then begin to chatter to other neurons running up the spinal column that an erection is in progress, and the brain is eventually informed of the matter. Thus arousal, which may begin like a knee-jerk reaction, is soon brought into our conscious awareness. Once again, there are aspects of these feelings that appear to happen to us.

In females, a similar blood-engorged response occurs in the clitoris. The clitoris is a sexually responsive knob-like structure. It too is composed of

How sex hormones work

Two hormones deeply involved in human sexual arousal are testosterone and estrogen. Here's what they look like and how they work.

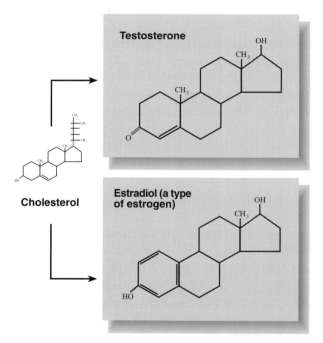

SEX HORMONES
Both testosterone and estrogen belong to a class of molecules known as steroids. Shown on the left, these sex hormones are simple molecules somewhat reminiscent of honeycombs. Testosterone and estrogen are remarkably similar in structure; indeed, they are created from the same precursor molecule cholesterol. Moreover, they can be interconverted (males, for example, possess an enzyme known as aromatase, which can convert testosterone into estradiol!).

HORMONES TRAVEL
Testosterone and estrogen play key roles in sexual arousal, working in regions of the brain such as the hypothalamus. In adults, sex hormones can travel great distances to exert their effects. Estrogen, for example, is made in the ovaries. To get to the hypothalamus, it must use the bloodstream (usually accompanied by some kind of carrier molecule). Once estrogen arrives at its destination in the hypothalamus, it must enter certain neurons in the organ to create its effects. How estrogen (estradiol, in this case) gets into neurons, and what it does once it arrives, is shown on the next page. The process, which involves binding to a receptor, can be divided into five overall steps.

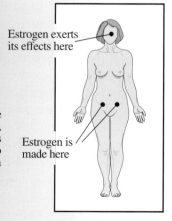

Estrogen exerts its effects here

Estrogen is made here

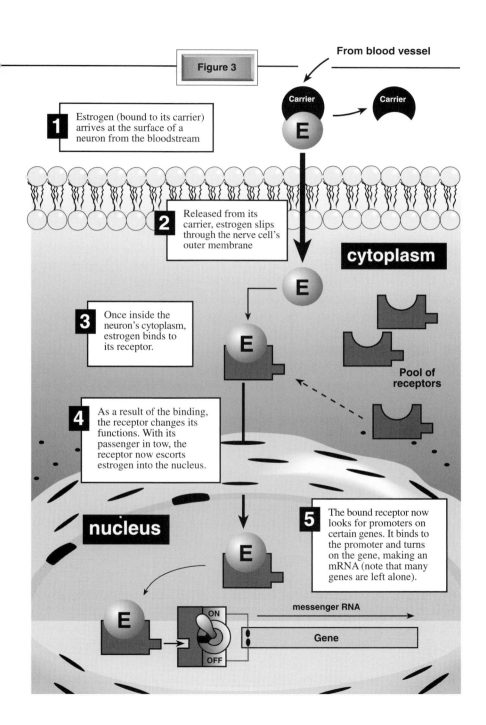

Figure 3

From blood vessel

Carrier

Carrier

E

1 Estrogen (bound to its carrier) arrives at the surface of a neuron from the bloodstream

2 Released from its carrier, estrogen slips through the nerve cell's outer membrane

cytoplasm

E

E

3 Once inside the neuron's cytoplasm, estrogen binds to its receptor.

E

Pool of receptors

4 As a result of the binding, the receptor changes its functions. With its passenger in tow, the receptor now escorts estrogen into the nucleus.

nucleus

E

5 The bound receptor now looks for promoters on certain genes. It binds to the promoter and turns on the gene, making an mRNA (note that many genes are left alone).

messenger RNA

E

ON

OFF

Gene

cylindrical structures (two of them) also running along its insides. These structures are called cavernous bodies. Similar to an erection, blood also flows into the cavernous bodies, causing the clitoris to swell during sexual arousal. Other parts of female genitalia also experience blood engorgement, including the vulva and certain internal tissues surrounding the vagina. Such an increase in blood pressure is a signal for the vagina to add lubrication to its outer walls and even to remodel its three-dimensional architecture, allowing for better penetration. As with males, various neurons within the female genitalia become stimulated during arousal, informing the brain of the autonomic nervous system's experiences.

From a cell's point of view, all this communication occurs over extremely large distances, between far-flung organs, between disparate kinds of tissues. That means a tremendous amount of information must be transferred – and quickly – in order for a coordinated response to be achieved. With all this required trafficking, you might suspect that the body uses more than just one set of tissues to achieve sexual communication. And that's a good suspicion. Another system, called the endocrine system, is also exploited. This system's highways are not neural, however, but circulatory in nature, exploiting not electricity but famous chemical couriers usually referred to as hormones. These hormones are created by specific glands, called, not surprisingly, endocrine glands. (The term endocrine is a catch-all word describing glands without ducts, yet still capable of secreting hormones into the circulatory system. Our brain, for some curiously good reason, is sometimes referred to as an endocrine organ.)

Endocrine system

The endocrine system comprises a series of glands whose job it is to dump hormones into the circulatory system, thereby communicating to the brain and other tissues within the body specific pieces of data, including sexual information. Two of the most well known hormones involved in human sexuality reside in a class of chemicals called steroids (see Figure 3). They are the very familiar testosterone and estrogen. (A side note: testosterone can mean any one of a family of related hormones usually called andro-gene hormones. I will use the word testosterone for simplicity's sake alone.) To describe how they work with sexual feelings, we will first talk about certain regions of the brain that help mediate sexual arousal. Then we will return to the hormones, and the genes that make them, discussing how they work with these regions to create arousal.

Hormones and the brain

As previously mentioned, there are at least four different physiological systems in our bodies that contribute to the feelings of sexual arousal. We have the nervous system, the genitourinary tract, the endocrine system and the circulatory system – literally millions of cells – all working in concert to give us our experience of the "poison of Venus".

Even if it is not always the first to know that feelings are being generated, the brain is the central clearing-house for most of the sexual information being processed by the body. But so far, we have not discussed the regions in the brain to which sexual feelings report. Exactly what area of our head is responsible for these feelings? How does it generate desire in response to the biochemical changes accompanying arousal? As you might suspect, there is no one part of the brain that generates the feelings of arousal, though there are a few major players.

In a few moments we are going to discuss areas of the brain that have odd sounding names like "nuclear accumbens", "anterior hypothalamus", and "ventromedial nucleus". Don't be too put off by these strange terms. Though complex, brain anatomy at one level is fairly easy to grasp. Brains have a top, which is near our scalps, a bottom, which is near our necks, and a mid-region which fills the space in between. The top areas, logically referred to as our higher regions, possess neurons that process sensory information, abstract thought, memory, and learning. The bottom areas, often referred to as the lower regions, are not nearly so sophisticated. They process primarily nonabstract reflexive behaviors, certain so-called vegetative functions (temperature regulation and digestion, for example) and distinct hormonal activities. The middle communicates with both regions and possesses a variety of higher and lower functions.

Embedded like fruit in gelatine within these top, bottom and middle regions are various structures possessing those funny names we just described. These structures are not isolated islands, but are rather all wired to each other, continually talking and cross-talking like excited teenagers on a telephone. At least eleven different structures mediate feelings of sexual arousal, and they are found in all three layers (see Figure 4). With such complexity, how do we know which ones are involved in sexual feelings? The data have come to us from a variety of sources, most notably surgical techniques and specific effects of drugs on sexual drive. To understand what I mean, consider an interesting (if painful) medical procedure practised since the dawn of time and perfected in the Middle Ages.

The procedure to which I refer is called skull trepanning. In Dante's time,

Lust and the human brain

Many regions of our brain work in concert to produce sexual feelings. Illustrated below are a few of these areas, with short descriptions about how they work.

A WORD ABOUT DIVISIONS

As you examine the various parts of the brain illustrated here, it will be tempting to view them as isolated islands, each performing some function independent of the rest of the organ. Indeed, this drawing, like those in many textbooks, provides brain "maps", seeming to cut the brain into discrete, function-only areas. In reality, such isolation could not be further from the truth. The human brain is a marvelous study not of insular regions, but of complex *connectivity* between vast numbers of nerve cells. These cellular cables hook up and criss-cross the regions mentioned here with such complexity that it can be difficult to separate a specific region from the connections. Indeed, many of the connections may be the brain "regions" themselves, responsible for functions of which we are currently only dimly aware.

You may also notice that many of these regions possess strange names. Researchers sometimes appear to enjoy naming various structures with words very few people pronounce correctly. Don't let the unfamiliarity put you off; experientially, you are well acquainted with the various areas in this drawing. Indeed, you are using some of the regions right now to comprehend this sentence!

CEREBRAL CORTEX

Helps to process sensory and cognitive data; aids in the interpretation of these data. The *frontal cortex* helps mediate storage of past sexual experiences, guiding choices of partners, relationships, specific learned sexual strategies, etc.

the mighty hypothalamus

Ventromedial nucleus

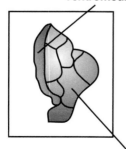

No organ may be more closely associated with human sexual responses than the hypothalamus, a smallish multi-nuclear organ shown on the next page. The hypothalamus is divided into several regions, such as the *medial preoptic nucleus* and the *ventromedial nucleus* shown on the left. These two regions help mediate sexual responses, the former with active sexual behavior, the latter with passive sexual behavior (see text).

Medial preoptic nucleus

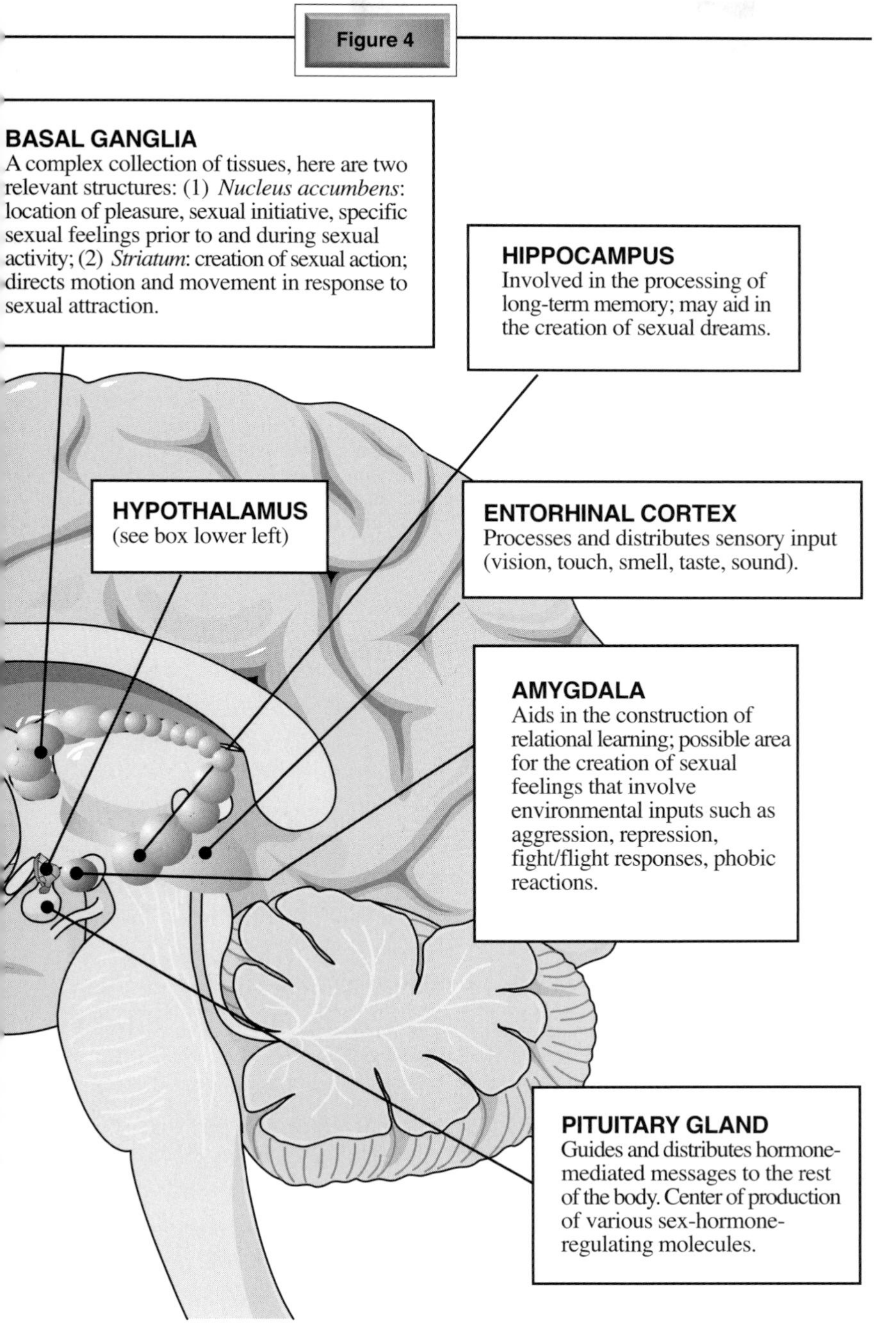

Figure 4

BASAL GANGLIA
A complex collection of tissues, here are two relevant structures: (1) *Nucleus accumbens*: location of pleasure, sexual initiative, specific sexual feelings prior to and during sexual activity; (2) *Striatum*: creation of sexual action; directs motion and movement in response to sexual attraction.

HIPPOCAMPUS
Involved in the processing of long-term memory; may aid in the creation of sexual dreams.

HYPOTHALAMUS
(see box lower left)

ENTORHINAL CORTEX
Processes and distributes sensory input (vision, touch, smell, taste, sound).

AMYGDALA
Aids in the construction of relational learning; possible area for the creation of sexual feelings that involve environmental inputs such as aggression, repression, fight/flight responses, phobic reactions.

PITUITARY GLAND
Guides and distributes hormone-mediated messages to the rest of the body. Center of production of various sex-hormone-regulating molecules.

it was common to think of pathological processes, emotional disorders or even bad temperament in terms of spirit possession. To drive out the evil spirit, it was necessary to create a hole in one's head, sometimes even putting salt or acid in the wound. This procedure, called skull trepanning, was always painful and quite often fatal. As awful a procedure as skull trepanning was, the suffering experienced was quite selective. Pain could only be felt by the poor wretch undergoing the procedure in the scalp and surface tissues. If the drilling or the chemicals penetrated the exposed brain tissue, and of course it did, there would mercifully be no uncomfortable sensation. There is a biological reason for this phenomenon. Though our brains are composed of many different kinds of nerve cells, there are no pain-sensing neurons embedded deep in our skulls. The head pain we do experience, even headaches, is because other tissues in our head are being stimulated.

This curious quirk of nature has been a friend to researchers interested in how the human brain works, including those interested in understanding sexual arousal. In the 1960s and early 1970s, researchers began opening up the brains of living patients, but only giving them local (cranial cap) rather than general anesthetics. This allowed the patients to remain wide awake during surgery, which means they could literally talk to the surgeons in the operating room. Such interactions proved invaluable for determining what areas of the brain were responsible for which functions. The surgeons began probing the deep structures of the brain with electrodes, stimulating the neurons in various areas, and then asking the patients what they were feeling. Because the patients were awake, they could answer the doctors' questions and, as a result, so-called brain maps were created. These maps detail the response of the patients, identifying where various sensations, feelings, and memories reside. One of the sensations researchers examined was sexual arousal.

A scientist named R.G. Heath noticed that when he stimulated an area of the brain known as the limbic septum (an area in the middle to upper region of the brain), his patients experienced pleasurable sensations. These sensations included extraordinarily intense sexual feelings, with the patients often experiencing orgasm. When Heath directly stimulated specific regions of this septum, the pleasurable responses were accompanied by "ecstatic" feelings of sexual arousal of such intensity that his patients often tried masturbating on the operating table. Both the reaction and the intensity were experienced by males and females. Other areas of the brain, including a thimble-sized structure known as the hypothalamus, could also mediate pleasurable activities. Could these regions, or perhaps regions closely associated with them, mediate sexual arousal?

Heath's work pointed the way for many researchers to take an active look at the brain and sexual response. As the years went by, the areas of the brain mediating sexual responses became clearer and broader in scope. Many years have passed since these studies were performed, and, as mentioned previously, many different regions of the brain have been identified as being part of the arousal process. While incomplete, a description of some of the regions responsible for sexual arousal is shown here.

Memory An area of the brain known as the frontal cortex helps in the storage and processing of memories of past sexual encounters. This area also helps guide sexual choices and the learning involved in creating sexual/relational strategies. As you might suspect, memory plays a great role in the human's experience of arousal; it probably does so for all animals.

Sensation interpretation As sensations arrive from the body, the brain must process, interpret and ultimately create cognitive information about what is occurring. Then it must decide what to do with the input being received. This happens in an area of the brain known as the cerebral cortex (the routing and processing of this sensory input also springs from an area known as the entorhinal cortex).

Sexual behavior As arousal proceeds, sexual behavior is generated. Heath also explored the areas in the brain that mediate this series of actions. Most famous is that tiny hypothalamus. Though small, this collection of neurons is involved in an equally exceptional number of vital functions besides sex, including temperature regulation, eating and drinking (we'll be discussing the various roles of this talented organ in nearly every chapter). Many researchers now believe that this organ has a critical role to play, perhaps *the* critical role, in both our arousal and continuing sexual activities. It has even been linked to sexual orientation. As researchers are intensely interested in the role of the hypothalamus, our attention now turns to its structure and function. Once explored, we will use this information as a gateway to consider the world of hormones and genes in sexual arousal, completing our discussion of this part of The Genetic Inferno.

The following important fact must be kept in mind as we discuss specific neural regions, however: they are all connected to each other. In fact, neurons are often said to "project" from one area into another, which among other things allows the regions to keep close tabs on their neighbors.

A person might make a central cognitive evaluation from the input he or she is receiving, but this evaluation comes as a result of the communication of various areas of the brain to one another over a period of time (the result of the projections, if you will). A whole palette of essentially adaptive responses can occur in response to these stimuli and can do so because of these connections. Even though the hypothalamus is one thimble-sized region, it must be remembered that it is not isolated from other brain regions.

Focusing on the hypothalamus

So let's discuss the biology of the hypothalamus and sex. To do so, I would first like to describe a very interesting meeting that occurred as Dante and his friends continued to explore the heated world of the Seventh Cornice. The meeting has to do with what he calls The Sodomites, and illustrates very nicely certain functions of the hypothalamus. It occurred about midway through his tour of the Seventh Cornice.

As you know, Dante and his friends are walking along the side of a mountain on a narrow trail, looking to the left at ghosts in the flaming fire. But these ghosts are not just milling about in the flames, writhing and scalding in sexual misery. Rather, Dante notices that they are walking in columns in a particular direction, a hopeful, seemingly natural direction, east to west, towards a setting sun.

As we discussed previously, some of these shades try to speak to Dante and his friends. But they are interrupted because, all of a sudden, another column of burning, miserable people comes marching forward from the opposite direction. In contrast to those shades talking to Dante, this column is composed of The Sodomites, homosexuals and sexual sinners not of the heterosexual variety. By not following the "natural" way, their march in the opposite direction is a symbol of their sin.

As this fresh column marches past the first, Dante witnesses an extraordinary thing. The Sodomites come up to members of the first column, and, without breaking stride, briefly kiss them. Dante describes it this way

> *Ants, in their dark ranks, meet exactly so, rubbing each other's noses, to ask perhaps what luck they've had, or which way they should go.*

After the kiss, the Sodomites shout, "Sodom and Gomorrah" and the heterosexuals reply with "Pasiphae", a Greek reference to bestiality. The columns then quickly part ways, with The Sodomites disappearing into the distance. Dante, as you might expect, is stunned by this odd interaction, and asks the people in the first column to explain it. While the purpose of our chapter is not to expound on thirteenth century views of sexual morality, Dante learns that this interesting meeting between hetero- and homosexuals is filled with symbolism and intrigue. For our purposes, there are two points worth noting here. First, the opposing columns met at a single point, and second there were issues of types of sexuality, even of orientation, when they interacted. The joining of these columns is clearly a study in the concourse of the antithetical.

It is just this idea of contrariness that produces a useful metaphor in our discussion of the hypothalamus and its role in sexual arousal. This important, tiny little organ is a region of the brain that also brings opposite sexual biochemistries together at a single point, just like the marching columns of the Seventh Cornice. Moreover, some researchers suspect that this region may also hold keys to sexual orientation. While the notion of aim is quite controversial, the fact of this organ's involvement in human sexual arousal is not. Let me explain this idea of opposites, by first talking a bit about the anatomy of the hypothalamus.

Structure

As we discussed, the hypothalamus is a tiny little region that sets up shop near the base of the brain. It is composed of front-end areas, one of which is called (here comes yet another neurobiological term) the medial preoptic nucleus. The organ also has back-end regions, one of which is termed the ventromedial nucleus. (We have actually run across the word "nucleus" before. In neuroanatomy, the word nucleus refers to a group of cells whose function can be identified from its physical structure and placement within the brain.)

Historically, there is evidence that part of the front end of the hypothalamus plays a role in generating male-specific sexual behavior. Such behavior has often been termed "active" sexual behavior, or "masculine" sexual behavior. The neurons in the front end appear to respond to particular sets of hormones and react in novel ways depending upon the type of hormone encountered. Conversely, part of the back end of the hypothalamus appears to play a role in the so-called female-specific sexual behaviors. This type of behavior has also been termed "receptive" or, perhaps more prejudicially, "passive" behavior. The neurons in this back end also appear to respond to particular sets of hormones, different from those to which the front neurons respond. Like the ridge in the Seventh Cornice, the hypothalamus is an area where opposite sets of sexual behaviors come together in one location. They undoubtedly interact, though there is evidence of specialization. In theory, your gender determines which region becomes predominantly active. The nerves of the hypothalamus, in this view, are the blank glass plates upon which the picture of heterosexuality is painted: males make use of the front of the hypothalamus to become attracted to females, and females the back to become attracted to males. I say the words "in theory" as this explanation is greatly simplified, and, even given a complete explanation, remains controversial.

The idea of gender assignments, neurons and arousal brings us to the other issue in that singular location on The Seventh Cornice: homosexuality. Some researchers theorize that homosexual activity arises because of confusion about which parts of the hypothalamus are activated in which gender during arousal. In this view, anatomical and biochemical changes in this organ help "decide" for the person what their sexual orientation is going to be. Thus, homosexuality may be partially explained as differential changes within specific neural structures during development of the human embryo. Those researchers who subscribe to this view often use the words "masculinization" or "feminization" for parts of the brain when describing the targets of sexual feelings. As you might suspect, there is a fair amount of contention associated with this assertion. It is not only the job assignments that are being actively debated at this point, but also the ability of environmental influences to either interact or even shape specific brain structures after birth.

A number of theories

Considering the knowledge vacuum, you might suspect that a number of theories have arisen regarding the role of the hypothalamus in sexual arousal. That suspicion is correct. Some researchers believe that the hypothalamus plays its primary sexual role not in arousal and signaling, but in behaviors that come afterwards, such as mounting and thrusting. Others believe that the organ plays its most important role by both creating the arousal experience and giving the arousal direction. With these two contrary ideas shouting at each other, it is clear that the jury is still out regarding the role of the hypothalamus in feelings of sexual desire.

It is, of course, a very simplistic notion to divide the labor of such a complex idea as sexual arousal into front and back ends of a single organ. As we have discussed *ad nauseam*, there are many interesting regions responsible for feelings of desire, the hypothalamus being only one element. You recall that Heath actually got the most intense sexual reactions from his patients by stimulating regions such as the limbic septum, which is a region separate from the hypothalamus. If the hypothalamus is involved, at some point it will need to talk to the septum, and that means they have to be connected. More recent data show that there do indeed appear to be vital and active connections between these two important regions of the brain. In fact, the hypothalamus is a virtual motherboard of neural connections to most of the eleven areas in the brain we discussed as having roles in arousal and desire. Many of these connections remain uncharacterized.

Having allowed for all these exceptions, we are not left completely in the neural darkness, marching our scientific theories away from a conclusive setting sun. Actually, there has never been a more exciting time to be in neurobiology than the present. We have been extracting data from some very interesting experiments in the tiny world of molecules, which may in the end shed light on many of the ambiguities of the different cell types. The last pigments I want to add to our biological painting of human sexual arousal come from characterizing desire in its most intimate detail – diminutive hormones, the little genes that make them and the smallish cells upon which they act.

Of cells, hormones and genes

We are now going to move away from the generalities of armies of systems and battalions of subordinate tissues. While it is extremely useful to categorize their biochemical reactions, it is also fairly frustrating to talk about them because so little is known about their interactions in humans. Yet we have enough data to at least make a comment on the first of the two questions we posed at the beginning of this chapter. We asked, "what part of the brain is responsible for sexual arousal?" The answer is that there is no *one* part, but rather many parts of the brain responsible for sexual behavior in humans. Now we are going to address the second question, which is, "what gene is responsible for sexual arousal in humans?" As you might suspect, the answer to that question again is that there is no *one* gene, but rather many genes responsible for sexual behavior in humans.

To talk about the functions of cells and genes and chromosomes is to enter a world almost as foreign as the netherworld Dante entered on his way to paradise. Before we begin, we need to discuss a few basic facts about

Structure and function of a cell

All human cells possess certain structures and functions in common. These include the presence of a nucleus, a cytoplasm, and a genetic code. Here's how they work.

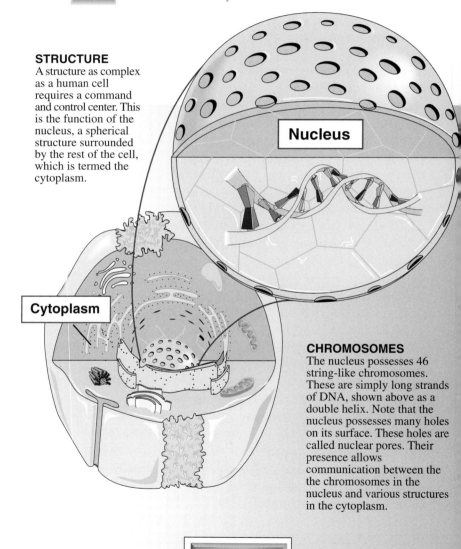

STRUCTURE
A structure as complex as a human cell requires a command and control center. This is the function of the nucleus, a spherical structure surrounded by the rest of the cell, which is termed the cytoplasm.

Nucleus

Cytoplasm

CHROMOSOMES
The nucleus possesses 46 string-like chromosomes. These are simply long strands of DNA, shown above as a double helix. Note that the nucleus possesses many holes on its surface. These holes are called nuclear pores. Their presence allows communication between the the chromosomes in the nucleus and various structures in the cytoplasm.

Figure 5

SOLVING AN IMPORTANT PROBLEM

Chromosomes are divided into discrete activatable regions termed genes. As discussed in the text, genes encode the information necessary to make proteins. There is a communications problem between the genes and the cell, however, as illustrated below.

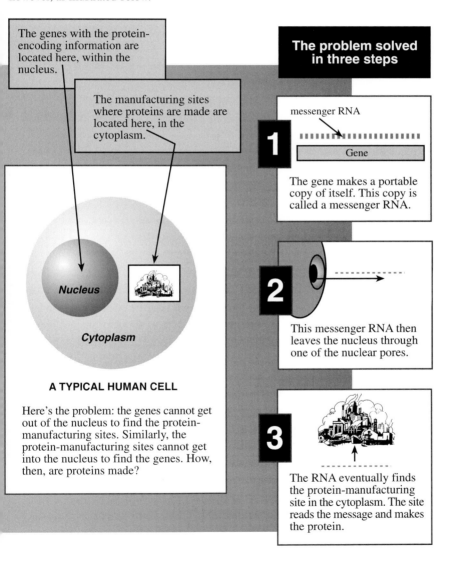

The genes with the protein-encoding information are located here, within the nucleus.

The manufacturing sites where proteins are made are located here, in the cytoplasm.

Nucleus

Cytoplasm

A TYPICAL HUMAN CELL

Here's the problem: the genes cannot get out of the nucleus to find the protein-manufacturing sites. Similarly, the protein-manufacturing sites cannot get into the nucleus to find the genes. How, then, are proteins made?

The problem solved in three steps

1

messenger RNA

Gene

The gene makes a portable copy of itself. This copy is called a messenger RNA.

2

This messenger RNA then leaves the nucleus through one of the nuclear pores.

3

The RNA eventually finds the protein-manufacturing site in the cytoplasm. The site reads the message and makes the protein.

the structure of human cells, and the function of the genes and chromosomes that reside within them (Figure 5). A typical human cell has a
nucleus, and the cell that surrounds the nucleus is under its control. This
surrounding area is called the cytoplasm of the cell. I want to explore some
of the biological structure and function of both the nucleus and the
cytoplasm, and use it to talk about the hormones of arousal.

The nucleus contains the command and control functions for the entire
cell, which as you know are embedded in structures termed chromosomes.
In humans, there are forty-six of these chromosomes. These chromosomes
are made out of a familiar material DNA, short for the term deoxyribonucleic acid.

A typical human chromosome can be further subdivided into regions,
based on their ability to perform (or not perform) certain functions. There
are, for example, regions that are very inactivatable and cannot do anything. Conversely, there are regions that can potentially perform some very
vital functions, and are said to be activatable as a result. (Activatable will
be explained in a moment.) The inactivatable regions we call "junk DNA"
– that's actually the scientific term! – and the activatable regions we call
genes. These genes, because they are part of a chromosome, are simply little chunks of DNA.

So just what is the function of a gene? It all revolves around the word
"activatable". The purpose of a certain class of gene is to encode the information necessary to make a protein. That may not sound like a big deal –
you can find proteins in the common muscles of that fatted calf – yet without proteins a human being would literally cease to exist. Proteins give our
bodies structure and form, and are the substance of our hair, our fingernails and, like fatted calves, our muscles. Without proteins, we would look
like misshapen blobs. Like genes, proteins are made of individual subunits,
here termed amino acids.

Providing structure is not all proteins do. Proteins are also involved in
the subtle tasks of the body, performing useful chemical reactions like
thinking and digestion. They are also vitally involved in the generation of
our feelings of sexual arousal. Proteins that perform work (rather than just
serve a structural function) are given a special name. We call them enzymes.
One of the functions of a gene, then, is to encode the information necessary
to make proteins that give us our structure. The other is to encode the
information to create these working enzymes.

Note that I said "encode the information necessary . . ." That was not a
slip of the tongue. A chromosome is wholly incapable of making a protein
on its own. Like I said, it can only encode the information necessary to

make one. The manufacturing site that *could* make a protein actually lies outside the nucleus in the cytoplasm of the cell. And that creates a rather substantial problem. In the odd world of cellular biology, a gene is incapable of escaping the nucleus. That means it can never find, let alone exploit, the cell's protein-manufacturing site. The converse problem is just as frustrating. The molecules that make up the protein-manufacturing site are absolutely incapable of getting into the nucleus to find the gene. Thus the instructions and the assembly areas lie in impossibly different areas of the cellular landscape.

How is this communication problem solved? The nucleus decides to send a message. The gene actually makes a tiny copy of itself, a mobile chemical termed messenger RNA (short for ribonucleic acid), and then sends that messenger out of the nucleus. This messenger contains the instructions originally embedded in the gene. It specifies the order of the amino acids (and the cell has the choice between about twenty different kinds) a given protein will possess. This instruction copy is often called messenger RNA, or simply mRNA. Note, when a gene is called upon to make a copy of itself in this fashion, it is said to be activated. If the gene is not currently making a copy of itself, it is said to be inactivated. Don't be confused here. There are regions of the chromosome that can *never* be activated (remember that the formal name for these areas is "junk DNA"). But a gene can be sitting there, waiting to become active, and not be junk.

Returning to the message then, it is entirely possible to create messages capable of leaving the tiny world of the nucleus for a single reason, namely that the nucleus, working something like a medieval castle, has a door and a drawbridge. We call this little aperture the "nuclear pore". The messenger leaves the nuclear pore, and, after some effort, finds the protein-manufacturing site. Molecules within the site read the message and then, according to instructions, make the necessary protein. That's how the problem of immobility is solved. Even though the molecules that command and control the cell never leave the nucleus, their instructions are still carried out. A protein, under the guidance of the messenger, is made in the cytoplasm and then allowed to perform its function.

Can a gene do anything else?

It is good to have this communication worked out between the nucleus and cytoplasm in the manufacture of proteins of course. But aren't there molecules within the body that are not proteins? And if there are – how are these other molecules made? And what does all this have to do with sexual arousal?

The answer to the first question is yes, there certainly are other molecules in our bodies besides proteins. Take our bones, for instance, or our teeth, which are made primarily of minerals (we also have structures in us primarily composed of sugars and fats). As to the relevance to sexual arousal, there are many hormones that are deeply involved in mediating these feelings, and many of these hormones aren't proteins either, but are in a class of chemicals called steroids. (Though you may have heard about hormones many times before, hormones actually have a formal biological definition. A hormone is defined simply as a traveling biochemical secreted by the endocrine system that alters the function of the target at which it has been aimed.)

The question is obvious: if genes truly are the command and control center, how does something that only controls proteins make something like a steroid? Are there other mechanisms of coding information inside a cell's nucleus? The answer to that question is no. Genes are truly and singularly the control-freak barons in the cellular kingdom. If that's the case, are different proteins made in such a fashion that they cooperate to create fancy steroid sex hormones, as if they are on some kind of assembly line? The answer to that question is yes. Genes do indeed encode proteins that can act in a corporate fashion. How these steroids are created and the relevance to arousal are both discussed below.

The hormones in all of us

There are a surprising number of hormones involved in creating human sexual arousal (see Figure 4). The biochemistry of most of them and their unique contribution to feelings of desire are still being elucidated. Undoubtedly, a complete description of arousal biochemistry will lie in their interactions but, since those interactions are also unknown, that description will have to wait. There do appear to be some major players, however, and it is to these that we turn our attention.

Perhaps the most widely studied hormone in human arousal is testosterone, that amazing steroid usually associated with male virility and masculine aggression. I say usually because, like men, women can also make this interesting steroid. And, perhaps counterintuitively considering its press, it appears to play an important role in the sexual arousal experiences of both genders (although other steroid hormones, such as estrogen, play important roles in a woman's sex drive; more on that in a minute). Understanding what testosterone looks like and how it functions will obviously tell us a great deal about what happens at the molecular level during periods of sexual arousal.

Testosterone looks something like four rings of a honeycomb that have been cut with scissors and then reglued together badly (Figure 3). It can be made from several sources, and it is here where our assembly line discussion comes into play. Surprisingly, testosterone can be made from cholesterol, the bane of the vascular systems of middle-aged men. The ingredients of a fried egg can actually be worked on by an assembly-line enzyme (an enzyme, as you recall, is a protein encoded by a gene; that's how a gene can control the creation of a molecule that is not a protein). What comes out when the assembly-line enzyme is finished is bona fide testosterone. Curiously, testosterone can also be made into estrogen, and that is done by yet another assembly-line enzyme. If you are thinking that this system is frighteningly flexible considering the importance of gender, you are understanding what I am saying.

At puberty

What happens next? Regardless of the source of testosterone, human beings eventually feel sexual arousal. Since reproductively competent feelings begin with the onset of puberty, it is useful to begin our discussion of the role of testosterone there. We'll start with males by peering once again into the deep tissues of the brain, then moving to the level of the gene.

You might recall that we talked about the hypothalamus, thimble-sized and polymathic, possessed of a possible sexual division of labor in its front and back ends. It turns out that this interesting region of the brain is hooked up by blood vessels to an organ that in the late Middle Ages was known as the slime gland (it was thought that mucous coming from the nose had its origins in an area of the brain). We now know that this organ, which is referred to as the pituitary gland, has a very different function. The connections it shares with the hypothalamus allow it to communicate to the human's far-flung tissues of reproduction. This is what happens in a boy's brain and genitals during sexual maturity.

Males

Around the time of puberty, and for reasons that remain mysterious, a molecular alarm clock goes off in a young man's hypothalamus, alerting him that reproductive proficiency is coming soon. Immediately, a series of genes is activated whose only goal is to make a chemical called a releasing factor. This factor will tell the boy's pituitary gland the fecundous good news. Once the pituitary receives the signal, it creates two race-horse-like hormones, which are released to gallop free into the circulatory system. These hormones quickly travel to the boy's testicles and tell these hardy spheres two things: (1) start making sperm and (2) start making testosterone.

And just exactly how is testosterone made? As we discussed previously, the steroid can be made from good old cholesterol, most likely given to the boy's body by the things he eats. Once in the testicles, the cholesterol is loaded aboard a molecular assembly line. Several sets of proteins then go to work and, as we discussed, the cut-and-paste honeycomb circle of testosterone is made. It is then dumped back into the circulatory system.

From that moment onward, the boy's body will continually create and regulate the amount of testosterone flowing through his veins. There's quite a bit of steroid, especially when compared to females. A human male has between ten and twenty times as much of the stuff, depending on the study you read. How this works with the arousals he feels will be discussed in a moment, after we talk about testosterone and females.

Females

As mentioned previously, women can also make testosterone. They synthesize the steroid primarily in their ovaries, and use a similar assembly-line-like model to manufacture it. The level of testosterone in females varies greatly with the menstrual cycle and even with the time of day. Between menstrual phases, for example, testosterone levels can be 80% higher in the morning than in the evening! These differences collapse to 12% morning to evening as the month wears on and the menstrual cycle goes through its phases. The highest levels of testosterone are recorded during ovulation, the time when the ovary donates an egg to the woman's fallopian tube.

Even though both genders use testosterone for arousal, the previous paragraph demonstrates how extraordinarily different the hormonal backgrounds are between the genders. These differences must be taken into account when examining human arousal. In human females, the issue appears to be much more complex than just dropping a little testosterone into the veins and watching for libido. Certain physiologists tell us that there are at least two different kinds of arousal (only one of which is mediated by testosterone) in human females. We will consider these differences as we leave this section and move on to consider how sex steroids affect the nuclei of brain cells.

General comments about arousal

Though it seems simplistic to say it, an arousal that is going to be transformed into the emotion of sexual desire must ultimately find its targets in our brains. At the molecular level, part of the story involves elucidating which hormones select which areas in the brain as stimulation proceeds. It turns out that the targets and hormones are used in slightly different ways in each gender. As we have discussed, testosterone is a big player for both sexes, though males use it to the exclusion of other available hormones. As one physiologist has put it:

> Men, as far as we know, have one basic sex drive pattern, which is predominantly governed by testosterone.
>
> *(Crenshaw and Goldberg, 1996)*

The situation in women is a little more complicated, however. It appears that women have several sex drive patterns, governed in part by that carousel of cycling hormones coursing through their bodies every month. These same researchers go on to say

> Women have several sex drive patterns, which are governed by a multitude of cycling hormones. These patterns change from day to day, depending upon the particular combination of sex hormones circulating through the female system.
>
> *(ibid.)*

These researchers have lumped female arousal into two prominent patterns. The first they call the receptive sex drive. It appears to be driven by the hormone estrogen and may be responsible for a woman's desire for penetration. It may also govern her desire to have sex with a partner rather than to masturbate. The second is called the aggressive sex drive. In this case, the hormone testosterone is the mediating biochemical. It appears to promote a woman's desire for genital sex, including a desire for orgasm, and may promote masturbatory behavior in addition to a desire for intercourse.

It is important, as ever, not to oversimplify these comments into thinking that this is all there is to sexual arousal. Testosterone and estrogen are simply two major players. A complete description is an integrative picture of many of these biochemicals creating reactions simultaneously in many parts of the body, including several regions of the brain.

That leads to the next question: if testosterone and estrogen help mediate sexual arousal, what areas of the brain should we focus upon in an attempt to understand how they spread their biochemical fires? As we discussed previously, many researchers believe that the hypothalamus is one of the primary regions of the brain involved in sexual arousal. We can thus get a few of the important genetic insights into desire by asking, "do testosterone and estrogen interact with the hypothalamus? and, "if they do, what happens when the hormones arrive there?"

The answers to these questions lie in the world of the gene, a world we will explore now in greater detail. To start our discussion on this topic (and gain a more molecular understanding) we will need to make a few general comments about how cells interact with hormones, and what occurs to the brain when they do. These processes are complex enough to require an illustration, and to do that we will return to Dante's *Divine Comedy* and talk about how he came to Purgatory in the first place.

The ferry boatman and a singer

To get to the great cliffs of Purgatory, a soul must be escorted across a vast sea by ferry. The entity that performs the ferrying task glows like light, and is called The Angel Boatman. No soul has a choice in the matter of their destination, but is placed in Heaven, Hell or Purgatory by a judgement of a person's moral choices. Throughout *The Divine Comedy*, there is the idea of both inspection and selection of souls. Dante gives us several surprises as to who he thinks makes it into the various levels of the afterlife.

The story of *The Purgatorio* actually opens up with a silent and thoughtful Dante sitting on the shore next to the mountain, talking with his companions. All of a sudden he sees a bright light in the distance, which is the Boatman carrying another load of recently deceased passengers. These souls are dumped very near the place where Dante is standing, and one of the new arrivals actually recognizes Dante. It is a man named Casella, who was a vocalist and was apparently a very good friend of the writer in real life.

As you might expect, Dante becomes very animated by this recognition and attempts to hug his familiar acquaintance. Dante's arms go right through him, to Dante's complete surprise (Casella has to take some time out to explain that he is now dead, has been transformed into a ghost, and can no longer be hugged). Nevertheless, he can still sing. Dante, continuing his excitement, begs Casella to break forth in song. The musician obliges, an activity disapprovingly noticed by the entity that guards the entrance to Purgatory. The story ends with Dante and his companions scampering over the shore, onto the sides of the mountain, and entering Purgatory.

These physical elements of *The Purgatorio*, the placement of souls into a ferry, the arrival at the shore causing excitement and disapproving notice, usefully illustrate our discussion of hormones and certain molecular interactions. Understanding these reactions in turn gives us a better understanding of the way biochemists view sexual arousal.

A matter of receptors

There are many parts of the brain (and indeed many parts of the body) that utilize both sex steroids in their daily functions. The hypothalamus is one of those regions. Since these hormones are often generated in other parts of the body and are given to the brain, the hypothalamus must encounter these steroids from the outside. Thus, understanding how the hypothalamus uses sex steroids is as simple as knowing how its neurons recognize and then apprehend them. We can then ask questions about the genetics of activation and utilization. Unfortunately, that is not as simple as it sounds. To discuss how neural cells in the hypothalamus recognize sex steroids, we have to talk about a class of molecules known as receptors. These are proteins that exist both in the cell membrane and, for our discussion, directly beneath the membrane. Here's an example of how they work.

When a traveling hormone finds a target cell, it's a little bit like a passenger coming to a loading dock and asking to board a boat. As in Dante's story of the Angel Boatman, the traveling hormone can't just get on board any time it wants to. Rather there is a selection process, involving an inspection of the would-be passenger and a decision made by the cell to either allow or not allow entrance. The reason for this selection is simple. There are many traveling molecules coursing through our bodies, not all of which are good for every cell. To protect itself, the cell is forced to become picky, and is in fact programmed to allow some molecules inside and some not. How this pickiness is achieved is an important clue to understanding the role of the hypothalamus in sexual arousal.

Most cells have the molecular equivalent of the gated loading dock we previously mentioned (a single hypothalamic neuron can easily have four thousand of them for testosterone!). The gate can sit either directly on the surface of the membrane, or just below the membrane. These gates are the previously mentioned receptors. Like a departed soul attempting to get on a boat to the afterlife, the hormone arrives at the receptor, and attempts to bind to it, hoping to seek entrance. Some cells will allow estrogen and/or testosterone to come inside and perform some function and some will not. The presence of the different receptors controls the pickiness of a given cell. (It is really a shape-driven event; a given traveling molecule will bind to its receptor only if the receptor is a particular shape. A more fitting analogy might be a key inserting into a lock.)

So what happens next? The binding of a particular traveling molecule with a particular receptor is not a neutral event. When a traveling molecule lands on the proper receptor, the event is often communicated to the command-and-control center of the cell, which, as you recall, is its nucleus.

The neurons that populate the hypothalamus have receptors for both testosterone and estrogen. When a human being is becoming sexually aroused, the requisite hormones begin pumping into the circulatory system. These hormones then travel to the brain. The presence of these hormones can be recognized by the hypothalamus, in part because of the presence of these receptors. Let's see first what happens when testosterone encounters the neurons of the hypothalamus, and then examine estrogen.

Testosterone receptors

The receptors that the hypothalamus possesses for testosterone are of a very odd variety. They are called *intracellular* receptors. Why are they called that? These receptors do not bob up and down on the surface of the cell membrane, like many receptors do. Thus, they do not project to the outside of the cell. Rather, intracellular receptors exist directly below the cell membrane. If steroid hormones like testosterone want to bind to their respective receptor, they are going to have to pass through the greasy barrier of the neuron's membrane and attempt to find them.

Fortunately, testosterone has the ability to do exactly that, slipping through the membrane of the neuron like a ghost slipping through a wall.

Once inside, it immediately encounters this intracellular receptor and binds to it. After the binding, an amazing process occurs. Almost as if The Angel Boatman were picking up Dante's friend Casella and escorting him to Purgatory, the intracellular receptor now picks up testosterone and takes it to the nucleus. Once in the nucleus, another amazing thing occurs. The receptor, carrying its testosterone passenger, now surveys the genes inside the nucleus. The reason for the survey is that it is looking for something it recognizes. Testosterone bound to its receptor is actually able to recognize a number of silent genes sitting in the chromosomes, akin to Casella recognizing Dante on the shores of Purgatory. While these genes do not then suddenly burst into song, they do become activated in a very particular fashion. They start making those messenger RNAs we previously discussed (in the parlance of molecular biology, this is called transcriptional activation). As you recall, these new messenger RNAs are then sent back out into the cytoplasm, where they guide the construction of new proteins. It is these new proteins that help mediate the sexual experience. So it is that, by hormones actually binding to receptors, the hormones turn on new genes. This stimulation is one of the ingredients necessary to experience arousal.

Estrogen and its receptors

In females another hormone system can also be utilized in the creation of desire, and it is to this system that we briefly turn. This is estrogen, as previously mentioned, a word derived from the Greek word *estrus*, meaning heat, receptivity, excitement, and the English word *generate*. Like most hormone systems, estrogen is part of an intricately balanced hormonal feedback mechanism. If blood levels of estrogen fall too low in human females, the pituitary sends a signal to the ovaries informing them of the deficit. In response, the ovaries secrete more estrogen, restoring the balance. As we have discussed, estrogen is also a steroid, and looks remarkably similar to testosterone.

The role of estrogen in female sexuality is quite complex. The steroid can be both excitatory and inhibitory, and what role is played depends upon a kaleidoscope of variables. This complexity is observed even at the molecular level. For example, multiple types of estrogen receptors exist in the brain. There is the type that works like an intracellular receptor, just like the testosterone receptor mentioned previously. Estrogen slips through the membrane, binds to its receptor, and the two go off to the nucleus,

How to turn on a gene

Genes must become activated to work properly. This activation involves regions on the gene itself and can even involve molecules outside the cell. Here's what happens.

THE POWER OF REPRESSION

Every cell in a human being possesses the genetic blueprint of the entire person. Too strange to be true? A cheek cell possesses not only the information needed to make a cheek cell, but also the information needed to make a foot. A liver cell has the information to make a skull. Why then is a cheek cell always a cheek cell, a foot always a foot and a liver always a liver?

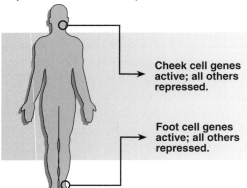

Cheek cell genes active; all others repressed.

Foot cell genes active; all others repressed.

The answer has to do with repression. As shown on the left, all the genes necessary to make a cheek cell are active in a cheek cell. But every other gene, like those foot genes, are rendered silent in a cheek cell. The same thing is true of cells in the foot; all the genes necessary to make a foot are active there, and all the others, including specialized cheek cell genes, are repressed.

HOW DOES A GENE LEARN TO KEEP SILENT?

As you might expect, knowing how to turn a gene on or off is an important part of a cell's daily activity. This regulation can be done because of the structure of genes. As shown below, all genes possess two overall parts: the part that encodes the information to make a protein and an on/off switch. The on/off switch is usually called the promoter of the gene.

If the on/off switch (promoter) is turned to off, the gene can no longer make a messenger RNA. It is said to be repressed.

If the promoter is turned to on, the gene can make a messenger RNA. It is said to be activated.

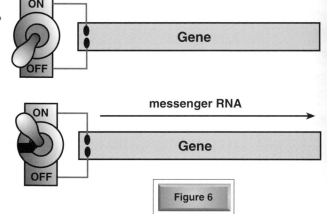

Figure 6

TURNING GENES ON BY USING OUTSIDE SIGNALS

The promoters of genes can be activated from molecules originating outside the cell. The molecules land on proteins found on the cell membrane (panel 1). These proteins are called receptors. Receptors can also be found directly under the membrane; the outside molecule has to pass through the membrane to bind to it (panel 2).

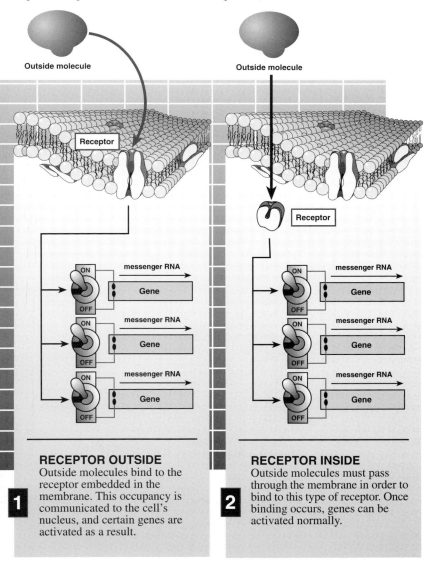

Outside molecule

Outside molecule

Receptor

Receptor

messenger RNA

Gene

ON
OFF

messenger RNA

Gene

ON
OFF

messenger RNA

Gene

ON
OFF

messenger RNA

Gene

ON
OFF

messenger RNA

Gene

ON
OFF

messenger RNA

Gene

ON
OFF

RECEPTOR OUTSIDE

1 Outside molecules bind to the receptor embedded in the membrane. This occupancy is communicated to the cell's nucleus, and certain genes are activated as a result.

RECEPTOR INSIDE

2 Outside molecules must pass through the membrane in order to bind to this type of receptor. Once binding occurs, genes can be activated normally.

The complex nature of lust

*An almost unbelievable number of tissues and biochemicals are involved
in the creation of human sexual arousal. Here are some of the systems
and molecules involved in the experience of lust.*

SYSTEMS
Besides the
obvious
reproductive
tissues, there are a
number of systems
involved in the
creation of sexual
feelings. Shown
on the right are
four of them.
These systems
work together in
an interlocking
and cooperative
fashion to mediate
arousal as well as
other sexual
functions.

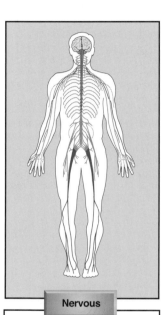

Nervous

The nervous system includes the
brain, spinal cord and peripheral
nerve cells throughout the body.

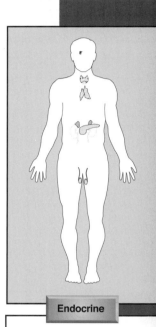

Endocrine

This system consists of discrete
ductless glands. These glands produce
various sex-related hormones.

SIXTEEN TONGUE-TWISTING MOLECULES
These four systems work with a variety of biochemicals to mediate
sexual arousal. Some of these molecules are listed on the right
(don't be put off by their big-sounding names; this chapter discusses
only two of the major players, testosterone and estrogen). This
list is displayed simply to illustrate how complex the molecular
nature of sexual desire is.

 Estrogen

 Testosterone

 Vasopressin

 Substance P

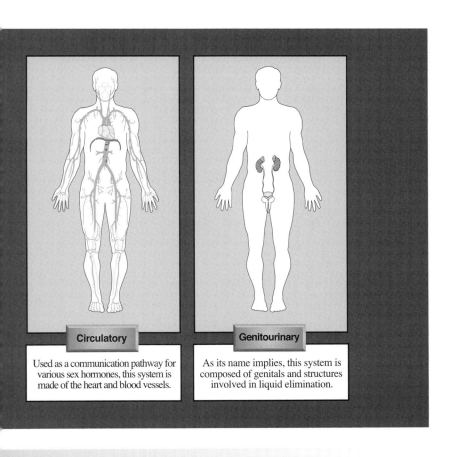

Figure 7

Circulatory

Used as a communication pathway for various sex hormones, this system is made of the heart and blood vessels.

Genitourinary

As its name implies, this system is composed of genitals and structures involved in liquid elimination.

 Dopamine

 Luteinizing-Hormone-Releasing Hormone

 Calcitonin Gene Related Peptide

 Oxytocin

 Dehydroepian-drosterone

 Prostaglandins

 Histamine

 Human Growth Hormone

 Vasoactive Intestinal Peptide

 Nitric Oxide

 Endothelium-Derived Relaxing Factor

 Various Excitatory Peptides

stimulating a variety of genes whose ultimate outcome is, in part, arousal. There is also an estrogen receptor that appears not to lurk underneath
the membrane, but is rather membrane bound, floating on a cell's surface,
as if it were a buoy floating on a sea (Figure 6).

Exactly how do these various receptors use estrogen to influence female
sexual arousal? Like the testosterone story, the real answer is that no one
knows. Estrogen receptors are found on nerves in the hypothalamus and
can directly influence the organ's neuronal responses. It is known that once
estrogen binds to its receptor, a series of biochemical steps is initiated
inside the neuron carrying the receptor. Soon, new genes are activated and,
as with testosterone, fresh proteins appear. Somewhere along the line, this
new activation is part of an overall reproductive emotional system, one by-
product of which may be sexual feelings. What proteins perform which
role in which neurons is an open question. How all of those neurons
interact with their new biochemical goodies to bring about lust is still a
mystery at this point.

Is that all we can say?

What then, can we say about lust at the molecular level? The same thing
we can say about it at the cellular level – it is the aggregate effort of many
players. There are specific sets of genes going off like firecrackers within a
defined set of neurons in response to a specific stimulus. It is as much a
system at the molecular level as it is at the level of cells, tissues or even
brain regions. In the case of biochemicals, literally hundreds of genes
are involved in the process. We have named only a couple, based around
the hormones testosterone and estrogen. In fact, there are many, many
more (see Figure 7).

The last sentence only addresses the biology of the experience, without
asking for the neurons involved in the psychological aspect of lust. I am,
of course, addressing the emotional components of the feelings of arousal. The problem is that the biochemical characteristics make much more
sense to a test-tube than do the emotional components. One has to tread
over the scientific no-man's land of mind and brain and consciousness to
get at how these emotions occur. As promised, we will briefly discuss these
three subjects, using arousal to inform us about some of the boundaries
of emotions with hormones. We will begin by tracing a history of minds
and brains, and then discuss our first important ingredient of consciousness, a function called long-term memory.

Beginning our history of mind and brain

The events in the Seventh Cornice reveal many interesting tidbits about medieval attitudes towards sex, from the process of conception to the ideas of perversion. Because of the drawn-out lecture Statius delivered as the poets ascended the staircase, we can organize these tidbits into some kind of framework. We have to be careful about the centuries, of course. When we moderns encounter Statius' words, we might immediately think of mind/brain issues. However, such contemporary divisions were not part of medieval thinking; therefore, the ideas behind Statius' monologue make an excellent starting point in our attempt to trace the history of mind/brain ideas. Let's take a look at what Statius said, discuss its place in real history (remembering that *The Purgatorio* is an allegory), and then talk about one component of consciousness: the presence of long-term memory.

The monologue of Statius can be divided into three parts. In the first, he talks about The Generative Principle. Statius states to our modern-day amazement that human males have something called perfect blood, a substance separate from normal blood that isn't even part of the body's own circulatory system. When perfect blood enters the heart of Dad, the blood changes into sperm. The sperm now flows to the genitals and, in the act of intercourse, is dripped over female blood. How does conception occur? Mostly because of masculine input (female blood, because it normally flows away in the menstrual cycle rather than taking some kind of form, is said to be "passive" in the process). This male blood is not an inert player, however, because it has passed through his heart and possesses what is called The Generative Principle. The upshot is that when the two bloods intermingle during intercourse, The Generative Principle allows the male blood to act something like a sculptor. It clots the female blood, working it over until it takes some kind of embryonic human form. Conception thus takes place.

The second part of Statius' comment has to do with acquiring a soul, an event that he thought occurred concomitantly with conception. The soul also originates from the male, arising because of that favored Generative Principle. Just as the physical embryo needs time to develop organs and tissues, Statius tells us that the soul needs time to develop certain mental characteristics. One of these characteristics is the ability to reason, a process in which God Himself takes an active role. The instant the brain becomes fully formed, God breathes into the soul a powerful spirit, unique to the person in the womb. This medieval kamikaze draws all life forces from within the body into a single responsible, unique, individually self-measuring, reasoning person. Statius actually compares the process to the role of the Sun in the manufacture of wine (though he says that wine occurs because of the co-mingling of the Sun's rays with the "sap" of the vine).

If the first two parts of Statius' lecture on the ascending staircase have to do with the beginnings of life, his final third is on its endings. He explains to his companions that, when death occurs, the soul separates from its body and falls of its own weight to one of several places. It can go to those tragic shores from which Hell may be entered. The soul can also go to a boat that is literally sitting on the river Tiber, waiting to take its passengers to Purgatory. The Generative Principle still resides in the soul, and shines out from it as rays of the Sun. This shining affects the very air of the afterworld, and just as The Generative Principle on Earth sculpted a body from menstrual blood, so The Generative Principle of the soul forms a ghostly body from the air through which it shines. Only this body isn't flesh, it's atmosphere, and Statius calls the new construction a "shade".

Influences of The Generative Principle

In Dante's world, we have super-active male blood and passive menstrual blood and souls and spirit-breathing gods and Generative Principles and boats and shades. All these ingredients eventually give rise to a reasoning, conscious, self-willed human being whose planetary lifespan is limited. Though it is clear that The Divine Comedy has many allegorical features, Statius' speech comes from real thirteenth century notions, such as a soul birthed within a body. From where did Statius get them?

Most history books tell us that the headwaters of the Western mind/brain dilemma started with the Greeks. Under the watchful influence of Plato, many Greeks also thought there was a soul trapped in an earthly body. The drives of that body, including sexual ones, were considered debased and carnal. Nevertheless, the soul was everything the body was not – immaterial, eternal, corporeal, and beautiful. Plato told us that only the soul could comprehend absolute truths and absolute aesthetics.

Though Plato's influence was widespread, it was not complete. One famous fellow, the philosopher Democritus, was called an atomist because he held that the entire world was made of very hard and quite uncuttable particles (the Greek word *atoma* literally means indivisible). "Entire" in Democritus' world also meant human thought, and included the idea of consciousness. There was no perfect soul with perfect beauty; Democritus felt that aesthetic qualities, and our ability to behold them, were simply subjective opinions supplied by the observer.

As you might suspect, the medieval mind pretty much ignored Democritus. By the time we get to Dante, the Biblical view of humankind is painted in very Platonic hues: a godlike soul seems to inhabit a fleshly body. Greek influences indeed saturate *The Purgatorio*, with many references to classical mythologies uttered by some very Christian mouths. With the entrancing influence of Greek body and soul ideas, the notion of mind and brain and consciousness as organizing concepts did not really exist in Dante's world. The medieval twist was to place the responsibility of the construct on the shoulders of a powerful God which was enforced by a similarly powerful church.

Statius' biological explanations do not hold up to the scrutiny of modern-day neurobiology, of course. Surprisingly, we have *not* after all these years been able to provide a full explanation as to the nature of mind and brain and consciousness. In fact, we currently can only create a palette of characteristics we know consciousness must contain, and simply guess at the larger picture. It is a little bit like looking at a stained-glass window that has been shattered into a million pieces; we cannot yet create the big picture even if we may be able to identify some of the colors used. And it is at this point that both Plato and Dante may inform us, for certain ideas common to the Greek and medieval ways of thinking show up in present-day twenty-first century notions. Let us consider one of the characteristics of consciousness, long-term memory, and see how its presence applied both to Dante's world and to our own.

Long-term ideas

As you recall from the introduction, we are attempting to illustrate that many subjective emotions – like sexual feelings – come about as an interaction between two processes in our brains. These two processes are: (1) an embedded emotion system and (2) consciousness. We have spent time discussing the first characteristic in terms of tissues and cells and genes. Long-term memory, our initial ingredient in the grand recipe of this grand idea of consciousness, will help us to understand the second component of sexual feelings, even if we have difficulty giving an overall view. Let us begin with an example.

Though he had no idea of its complexity, the ability to create the astoundingly rich poetry of *The Purgatorio* required many processes working coordinately (and often simultaneously) in Dante's brain. Each process required information stored in Dante's neurons in most cases many years before he wrote *The Divine Comedy*. The first line of Canto XXV, for example, mentions Taurus and Scorpio in relation to the time of day. From it we get the time the poets visited the Seventh Cornice, about two o'clock in the afternoon. Writing such words required of Dante a past working knowledge of the horoscope, the association of the horoscope to the Sun, how their configurations relate to knowing the time, and how all of this would be used to place the text in his description of the Seventh Cornice.

As we explore a little more, we see that even an individual concept, such as writing the word "Taurus", required a great deal of previously stored information. There is solid evidence that such nouns are filed in a human brain in association with many other memories, and may be stored in more than one place. As Dante wrote, he might visually be thinking of the constellation of Taurus, or of the somewhat dangerous bull chained to a fence in the marketplace, or the teacher that taught him to recognize stars. Literally thousands of associations could be possible with just the single six-letter word.

At a deeper level, Dante's ability to write those six letters came from memories first embedded in his brain as a Florentine school-boy. The individual letters are stored somewhere (there is evidence that the brain houses vowels and consonants in different places), the entire word is stored somewhere (perhaps as a graphical representation, consigned to a different area than the text), and the ability of his right hand to move in a particular way in response to the letters he wants to write is also stored somewhere. Even the idea that a well-inked pen when applied to paper may create a word had to have been packaged in some cerebral corner. Dante was constantly summoning an almost unbelievable constellation of memories in ever-changing combinations just to conjure up the first few words of Canto XXV.

What is the point of this description? Because writing the word "Taurus" required a tremendous amount of information drilled into his head in some cases *decades* before he wrote, the poet Dante had to have some kind of long-term storage device in his brain. Indeed, he had to use it all the time as he wrote *The Purgatorio*. Since writing is a process that involves consciousness (if Dante were in a coma he could not create poetry), one of the ingredients of consciousness must include that long-term memory storage device.

Our next question is then straightforward and direct: is there scientific evidence for the presence of a long-term storage memory device in our brains? The answer is an enthusiastic "yes". We not only know that such a mechanism exists, we also have a pretty good idea where parts of it are located. We are even beginning to get a handle on how some of the molecules involved cause us to remember items like "Taurus".

How do we know where it exists and how it works? We are going to talk in some detail about memory subsystems in future chapters. For now, suffice to say that techniques exist that allow us to examine parts of a living brain while the brain is performing certain tasks. Some of these techniques include watching a brain as it encodes and recalls facts. In order for facts to become permanent parts of our cerebral hard-drive, we have to coax them out of one area of our brain and into another. That coaxing requires changes at the level of brain cells, and also alterations in the genes deep inside the cells being recruited. Once the alteration has occurred, the fact becomes part of our long-term memory. We will leave to those future pages the detail about what brain cells are involved, and what molecules within them make the neurons behave in certain ways.

How does the fact of long-term memory bear on our ideas about consciousness? While the discovery of long-term memory does not alone define it, nor settle the mind/brain controversy, it does give us a solid science to stand on as we discuss individual ingredients. Both Greek and medieval ideas make use of long-term memory in issues of identity and awareness, even if they were worlds away from defining it (none of them could recall the mythologies, for example, without such a device in their skulls).

Long-term memory and sexual arousal

While it is useful to determine the ingredients of consciousness even if we have a hard time defining it, this is primarily a chapter on the biology of arousal. Before we leave our discussion of consciousness, however, I would like to discuss how the mechanism of long-term memory might associate with lust. In so doing, I hope to reveal an important point about human sexual desire and our brains. It turns out that there are many roles for long-term memory in human sexuality. One of the most important of these is the role of cultural influence.

Without going into great detail, the following sentence is pretty much neurological dogma: the human brain is structured in such a fashion that it can be deeply influenced – even reshaped – by specific environmental experiences. As an example, you may know people who sink into deep depressions if something horrible happens to them, changing mechanisms in the way neurons talk to each other *even at the level of the gene*. Since we are extremely social creatures, a great deal of our environmental input is derived from the human culture that surrounds us. And it influences us because we have the ability to remember what occurs to us over long stretches of time. Thus long-term memory plays one of the most important roles in our ability to maintain cultural traditions, and even our identities.

This is true even – perhaps especially – of sexual feelings. There are males, for example, who are deeply aroused by stiletto high-heeled women's shoes. The arousing influence is so strong that some will risk breaking the law, perhaps by stealing these shoes, in order to satisfy this odd fetish. True physical arousal accompanies the presentation of a twentieth century creation.

Fetishes are thought to occur in part because of a previous pairing of sexual feelings and/or experiences with a cultural object. Though there are neurological correlates to sexual addictions and fetishes, there is no region in the male brains hard-wired to obsess over twentieth century women's shoes. In other words, the culture is going to have to present the addict with the item in order for the obsession to find its focus. While the culture may not directly establish the fetish, the culture helps shape its complexion (in another era that did not have such shoes, the fetishist would be forced to focus on another culturally supplied object). Without the shoes, the addiction would have a different shape, and without long-term memory there could be no such pairing with sexual feelings. This is a way in which culture, arousal and memory can work together.

This fetish idea is only useful as an analogy, however. This is because the influence of culture on sexual arousal runs much deeper than just the choice of inanimate obsessions. You may recall we discussed that cultural influences cannot be easily separated from biological ones. For good or ill, sexuality is an important component of human identities, and such identities are rooted not only in neurons but also in the cultural normatives swirling around them. While the purpose of this book is not to provide the evidence that the previous sentence might demand, the best view of arousal might be a synthetic one: our biology presents us with specific sexual potentials, and culture and personal experience mold these potentials in specific ways. The link between this and our memories then becomes quite obvious: cultural influences come directly from our ability to recall certain events in time. We retain knowledge of these events, and their effects, because of long-term memory. This very important component of consciousness is intimately associated with sex, which is itself shaped by the culture that surrounds it.

A summary and a parting thought

We've been talking about a lot of molecules and tissues and souls and memory components here, so let's take a few minutes to put things into context.

As we discussed at the beginning of this chapter, we are working with the following model regarding emotional feelings: subjective emotions are experienced when two powerful collections of neurons within our bodies interact with one another. The first collection of neurons comprises our embedded emotion system. This has many biochemical roots, and we discussed two powerful hormones involved in the process, testosterone and estrogen.

The second collection of neurons is that group that comprises our conscious awareness. Though as surely rooted in our brains as the hypothalamic interactions, the biological basis behind consciousness is still so poorly understood that we are forced simply to list some of the important neural components without creating a big picture. The component mentioned in this chapter is long-term memory. One of the most critical contributions such storage makes to our understanding of arousal does not first seem very biological: the presence of a long-term memory system shows us how biology and social forces interact and shape each other. Such memory allows us to establish, recall and become influenced by the sexual elements of our own culture at particular points in time. This isn't trivial; sexual subsystems are determined by the direct and unique experiences of life, at both the personal and the social levels. Long-term memory allows the influence of culture to paint individual sexual feelings in very specific hues. It is a very important ingredient to understanding the nature of human consciousness.

To summarize then, the emotional experience of lust comes when it suddenly dawns on us that the sexual subsystem has been aroused. This interaction between these two battalions of cells is quite dynamic, and can be activated from a variety of sources. We can think some internal sexual thought, which then activates the sexual subsystem, which at some level is brought into our awareness, and then we experience the subjective emotion of sexual arousal. Conversely, something erotic can happen externally to us, which also activates the subsystem, which is then transmitted to our consciousness. This allows us to become aware of the feelings and once again experience the emotion.

The enormous number of tissues involved, and the fact that *awareness* of sexual feelings is involved somewhere in the collision of two large neural processes hints at something very important. I wish to discuss it before we leave this chapter.

Parting thoughts

When one considers the biology of sexual arousal, it becomes very clear that the generation of the feeling has a certain automaticity, almost to the point of uncontrollability. This automaticity implies that the emotion of sexual feeling, perhaps like all emotions, is a process that happens *to* us, rather than being under our absolute control. Consider, for example, how impossible it is to fake feelings of sexual arousal towards someone. Couples who are happily married (and have absolutely no intention of having sexual activity with anybody else) can nonetheless be aroused by persons outside the marriage. This odd presence means that the social test of fidelity is not *whether* the feelings occur or not, for they seem to come of their own accord. The test of fidelity is how one chooses to act on the feelings once they occur. In other words, the emotions truly do appear to be something that crashes into our existence – even if we object – and our freedom of will is restricted to choosing how we act once the feelings come. Quite automatic indeed.

There appear to be several reasons for this unruly situation. First, the neural connections in the brain that fasten our emotions to our more advanced thinking processes, the so-called cognitive systems (and I believe there is a difference between emotion systems and cognitive systems), appear to be much stronger than those that run from the cognitive systems back to our emotional ones. Second, there is a great deal of evidence suggesting that many types of emotional processing occur in the absence of conscious awareness. This includes sexual feelings. The emotional meaning of a sexual experience may begin to be evaluated and processed in certain regions of our brains before those regions that enable us to perceive those appraisals have been activated. They thus lie outside our conscious control. Simply put, subjective emotional feelings of lust may be constructed like an iceberg with most of the effective parts lying hidden beneath the surface.

That troubling fact leads to an even more troubling idea. . . . If it is true that a tremendous amount of lusty information about which we are only dimly aware is being processed in nether regions, then we really don't have any way of knowing what arousal actually looks like. In fact, there may be no such thing as a subjective feeling of lust at all. We can't look at brain location very comfortably, certainly. There is no one region in the brain that governs arousal, just as there is no one region that governs human emotion (nor is there just one overarching responsible bodily tissue, or gene sequence that gives us sexual feelings). There is just the presence of groups of tissues working together in a certain way to create some down-the-line evolutionary goal. Many of the tissues employed for sexual arousal also have day-jobs doing other things, which flies in the face of creating discrete task-only digestible brain functions. (For example, our eyes, our ears, even our skin are certainly integral parts of a system of sexual responsiveness. But that is only one aspect of the overall enormous burden they carry to extract sensory information from a hostile environment. Ditto with the hypothalamus, with the brain regions responsible for long-term memory, even for the genes involved in sexual responses.) The over-whelming biological lesson is this: it is not only the presence of the tissues themselves, but the way the tissues are used in combination that results in our perception of the emotion of sexual arousal. The ghost in the machine is a pattern, not an insular brain region or monolithic gene. When a sexual subsystem murmurs to our consciousness of intimate matters and subjective emotion results, it is a voice whispered by literally millions of molecules. And these millions in many cases, have other things to talk about.

This idea, if true, creates a hypothesis that flies in the face of convenient labels. There may be no such thing as the emotion of sexual arousal, or indeed any emotion. These feelings may just be the by-product of a clash of neurological systems working together in pursuit of survival. Please don't get me wrong here. Sexual subsystems certainly exist in the brain (an important part of the job description of evolution after all is to fling genes into future decades; there are distinct mechanisms built into our brains to make sure it happens). But the subjective sexual emotion of lust as we experience it may only be an artificial label humans apply to organize the feeling, with the label never describing something the brain really does. The subjective feeling may just arise from the collateral fallout of the enor-mous number of biochemical engines that must be started to ensure repro-duction into the next generation. Who knows? Lacking a more complete description of the iceberg, we have no way to tell anybody what its overall shape really looks like.

Taken together then, our first great lesson in attempting to understand the nature of emotions may be a nihilistic one. This need not be totally uncomfortable, for in many ways we experience such negative subjectivity on a daily basis. Take that 185-square-meter stained-glass window in the York Minster. When the Sun shone through it, I thought it was the most beautiful thing I had ever seen. An artist acquaintance standing next to me thought it was one of the gaudiest things he had ever seen. Who was right? From a strict reductionist point of view, there are no such things as atoms of beauty or molecules of gaudiness capable of being mixed into the paints or the glass or the plaster of the work. Once the cumulative end product of the sun shining through the window was experienced, there was simply a point of view.

CHAPTER THREE

Gluttony

"All those who sing while weeping in their pain
Once loved their stomach-sacs beyond all measure
Here, thirst and hunger wring them clear again."

-Canto XXIII, The Purgatorio

"Hungry?", the professor asked the class, as he pulled a golden peach out of his briefcase. Of course they were hungry, the class running in the hour just before lunch. The red-haired student, seated at the front of the class, seemed to eye the fruit with a special intensity, almost with an incandescent, impish quality. "Would you like to eat this peach, young man?" the professor asked him. As the student nodded enthusiastically, the professor handed the peach to his teaching assistant, who threw it to the student. Just before the young man caught it, the professor suddenly lunged at him, grabbing the peach in mid-air and shouting at the top of his lungs, "you shall not eat of the fruit!". The class exploded into laughter, while the professor, pulling back from the student, held the peach aloft like a trophy.

"There you have it, sir," the professor said, looking down at his unhappy charge. "That is what it is like to be an inmate in the Sixth Cornice". The young man, no longer looking mischievous, was glaring. "That's the level reserved for the gluttonous, where those who can't control their appetites. . .", and here he threw the peach back to the red-haired young man, ". . . are made permanently hungry to remember their sins". The student grabbed the fruit and quickly shoved it into his backpack.

As the laughter of the class died down, the professor began to explain the day's lecture. He was going to describe the hungry world of the Sixth Cornice, made especially for those people who on Earth could not push themselves away from the table. "The cornice was dominated by a couple of trees," the professor said, "one had fruit so fragrant and so delicious looking that the poets stopped in their tracks, staring almost in reverence. They could hear a voice coming from the center of the tree. It said, 'You shall not eat of the fruit'."

That's not the only voice they noticed, the professor went on to explain. As the poets continued to stare, they heard a chorus of mournful voices behind them, gradually becoming louder and louder. Eventually they saw the inmates of the Sixth Cornice in group formation, emaciated and wasting, running very fast, and practically skeletal. The outward appearance of the Gluttons reminded Dante of the ancient Greek character Erysichthon. "But there was nothing Greek about it," continued the professor, "It looked like more like a modern day aerobics class, appropriately enough, from hell. They were doomed to circle endlessly around the cornice, every time encountering the tree, every time hearing that obnoxious voice from the tree, not once being able to eat from the fruit."

Typical of the poets' sojourn in Purgatory, there was some interaction amongst themselves and also between them and the inmates. Statius gave another long lecture to his companions on the nature of human beings.

After the lecture, the poets met the running inmates, who were deeply surprised, even awed, to find a real live human being like Dante in their midst. And Dante recognized yet another shade as a friend, an old acquaintance named Forese, and stopped to chat. Forese helped identify several of the inmates in the aerobics class, including a Pope (whose sin was an overfondness for an Italian delicacy made of eels), and several other poets, one of whom gave Dante a prophecy. After a few more words, Forese dashed off to catch up with the rest of his jogging party. The journey eventually ended when Virgil, Statius, and Dante encountered one more tree, also heavily laden with fruit. It was Eden's tree of life. They saw a rather cruel sight, for this tree was surrounded by more inmates, begging with arms uplifted, all asking for a tasty bite. As with the first tree, their petitions were denied, but not before the inmates got yet another lecture on the sins of gluttony.

What we are going to discuss

In this chapter, we are going to explore some of the biology of the appetite these poor inmates suffered, focusing specifically on the desire for food consumption. For our purposes, hunger means the emotions springing from the purely biochemical reactions generated by energy needs. We shall start by describing interactions between brain and body, revisit the by-now-familiar hypothalamus, then explain some newly discovered genes involved in appetite generation.

With this biology under our belts, we shall return to our discussion of how biological inputs influence the generation of human feelings like hunger. We will do this by revisiting our history of minds and brains, talking about a concept called dualism and a man named Rene Descartes. Lastly, we shall see how our discussion of hunger influences our notions of consciousness, and add yet another ingredient as we seek to define exactly what it is.

Looking at the biology and then discussing mind and consciousness addresses both parts of the emotion model we have been following: an emotion system breaking into conscious awareness. This model describes humans; however, the data I am about to present were not discovered first in human studies, but rather in animal ones. That's okay as a starting point, and in fact leads us to an interesting evolutionary reflection. It is on this reflection that I will end the chapter. In some cases, our biochemistry is almost uncomfortably similar to that of other creatures. In other cases, however, the data show just how alone we really are, and that what applies to animals does not necessarily apply to us at all. Both of these ideas are important as we attempt to understand better the hungry souls of the Sixth Cornice. I will return to these ideas at the end of the chapter.

So let's get into the biology

With those promises of future comments in mind, let us dive into the biochemical world of energy consumption. You've probably already experienced the feeling today. "I'm hungry! Give me something to eat" is a powerful exclamation most of us utter in some form several times a day. I have to admit that I'm a sucker for babyback pork spareribs, especially the smoked kind marinated in a molasses-based barbecue sauce. Even as I'm typing out this sentence I can feel my hypothalamus screaming at me to stop writing this book, begging me to go down to the local barbecue and stuff my body full of smoked meat. As I contemplate what I'm feeling right now, the desire for food seems rather uncomplicated. Indeed, feelings of hunger seem to be some of the most familiar and straightforward that a human experiences: some food-related stimulus occurs, and we encounter a craving to eat. The origins of those feelings appear equally elementary. We might smell some wonderful concoction coming from the kitchen and become hungry for it, even if we weren't previously thinking gluttonous thoughts. We might think of a favorite food, such as those ribs, in the solitude of our thoughtlife, and our bodies respond to the dictates of an internally supplied daydream. Appetite appears to be a simple two-laned highway, as lust once appeared, one lane arising from the body and going to the brain, the other lane arising from the brain and going to the body.

Such observations can seem so straightforward that some people simplify the concept when they inquire of its origins. They may ask, "what region of the brain possesses the appetite highway?" or even more simply, "what is the gene responsible for feeling hungry?" As intuitively obvious as these questions may sound, appetite is actually much more complex than just a two-laned highway. There are a few major themes of physiological hunger, but there is great complexity in their development.

These biological themes consist of many interactive physiological systems, several regions of the brain, numerous biochemical mechanisms and hundreds of supporting genes. There is a *system* of tissues whose interaction gives us the feelings of appetite, but no one true region or one true gene is alone responsible for the response, not to mention the subjective emotion that derives from it.

Attempting to simplify the emotion of hunger is inaccurate for another, nonbiological, reason. As was true for lust, feelings of hunger cannot be divorced from their social contexts. Individual food preferences are marinated in geographical, social, attitudinal and, perhaps most important of all, economic variables. The end result is that different people groups in time and space end up experiencing their appetites in very different ways.

Another cultural component that cannot be ignored concerns something that might almost seem counterintuitive. It may seem elementary to you (it certainly seems to me) that we desire the rewards we find pleasurable, and find pleasurable the rewards we desire. When it comes to the generation of appetite however, this intuitive link can actually break down. In fact, there is some evidence that "wanting" and "liking" may be under the control of different (though linked) regions of the brain, and therefore separably defined. What determines our likes? What determines our wants? No one really knows. Researchers agree that the answer has powerful environmental ingredients, and, since we are talking about human behavior, we really mean cultural ingredients. These are experienced psychologically. While we are not going to dwell on the psychological components of gluttony in a chapter on gene biology, their interactions with these social variables cannot be conceptually ignored.

So what do we do?

Given the complexity of the task, can we *really* talk in a scientifically meaningful fashion about the nature of our appetites? The answer is yes, at least in part. We know a number of the major themes involved, even if we cannot yet write an entire score. The story of appetite generation is truly a story of what goes on in our heads, in our bodies and the communication that lies between them. We will begin our discussion with the brain, considering the function of a very familiar region, the hypothalamus. We will then move to a description of two tissues in the body that help keep the brain informed of the food supply, liver tissues and fat tissues. Finally, we will talk about certain recently discovered hormones that serve as messengers trafficking between our brains and our bodies, equally involved in alerting the brain about food requirements.

Studying research into appetite control can be very frustrating, and the reason is that the underlying basis for the feelings is quite simple, even if

Energy, food and appetite

Though the mechanisms are very complex, appetite is involved in the conceptually simple balance between energy expenditure and food intake. Here's one view of the overall framework.

OVERALL

STARTING WITH AFFERENT SIGNALS

1 As you know, the amount of food we consume gives our bodies a certain weight. The experience of appetite helps us to regulate our food intake, thus affecting our weight. There is a rather complicated series of processes controlling these simple ideas of energy expenditure and food consumption. This delicate caloric book-keeping is ultimately controlled by our brain, especially the hypothalamus. It receives sensory information from all over the body collectively termed "afferent signals", as shown below.

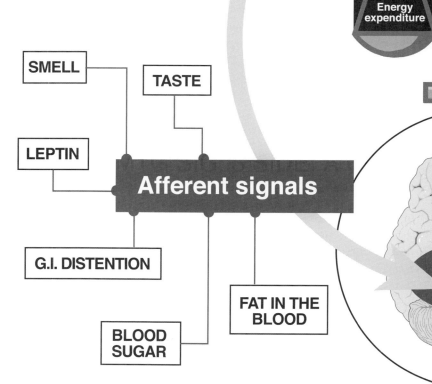

Energy expenditure

SMELL

TASTE

LEPTIN

Afferent signals

G.I. DISTENTION

FAT IN THE BLOOD

BLOOD SUGAR

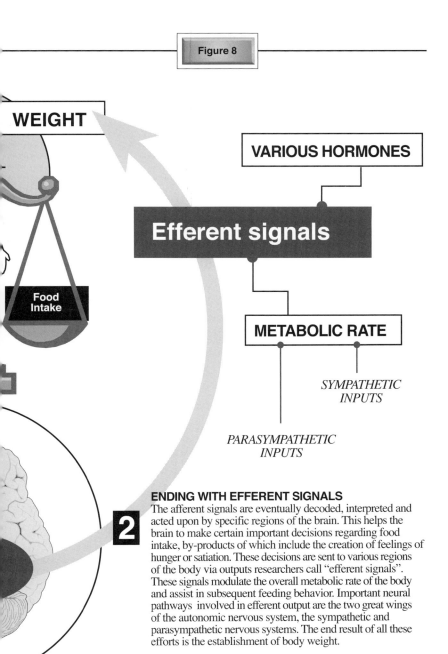

Figure 8

WEIGHT

VARIOUS HORMONES

Efferent signals

Food Intake

METABOLIC RATE

SYMPATHETIC INPUTS

PARASYMPATHETIC INPUTS

ENDING WITH EFFERENT SIGNALS

The afferent signals are eventually decoded, interpreted and acted upon by specific regions of the brain. This helps the brain to make certain important decisions regarding food intake, by-products of which include the creation of feelings of hunger or satiation. These decisions are sent to various regions of the body via outputs researchers call "efferent signals". These signals modulate the overall metabolic rate of the body and assist in subsequent feeding behavior. Important neural pathways involved in efferent output are the two great wings of the autonomic nervous system, the sympathetic and parasympathetic nervous systems. The end result of all these efforts is the establishment of body weight.

the mechanisms are not. If an organism has enough energy, the organism will live; if the organism does not have enough energy, the organism will die. This means that there has to be a balance between energy input and energy output, and ways to assess and respond to how the balance is tipping at any one time. Most researchers summarize the problem by saying that the control of eating is inversely proportional to the overall rate of fuel consumption. If that's true, it follows that certain trip wires must exist in the body telling the brain about the food supply. And indeed there are. The overall framework looks something like this (see Figure 8):

1. Our appetites, mediated in part by our brains, help us regulate our food intake. The amount we consume gives us a certain mass to lug around.
2. That mass is maintained over time by a rather complicated series of circuits controlling our food consumption and our energy expenditure. These circuits are really the trip wires.
3. This energy book-keeping is ultimately balanced in our brains, which receive sensory signals (often called afferent signals) informing the organ of the body's internal nutritional state and certain external cues, like the availability of foodstuffs.
4. These afferent signals are decoded, interpreted and acted upon by specific regions of the brain. As a result, the brain generates certain output signals (often called efferent signals) that control eating behaviors and the ultimate expenditure of energy. That, of course, brings us back to step 1.

Those afferent signals come from an astonishing variety of places. There are the obvious inputs, such as tasty little stimuli coming from the nose and tongue. However, there are other signals that inform the brain about hunger, such as mechanical distension of certain tissues in the gastrointestinal tract, certain molecules in the blood (like sugar and fatty acids), blood-borne hormones, even from communication with our liver.

The efferent signals also involve a large variety of tissues. Such signals ultimately control those two great wings of the autonomic nervous system we mentioned in the last chapter, the sympathetic and parasympathetic nervous systems. These in turn modulate that gross national product of the body, the metabolic rate. In cooperation with other hormones, such as thyroid hormone, these signals assist in controlling the subsequent feeding behavior.

Taken together, the brain and the body in energy balance are like two voices in a good medieval motet. They are saying separate things, are completely interactive, complicated, and exist to achieve a certain goal. We are now going to look at some of the biochemical themes of which this interaction is composed, starting with the brain, and see how they work together.

The brain

When considering the communication between head and body, the most solid thing anyone can say about the generation of appetite is this: it involves nerves. The second most solid thing anyone can say about appetite is this: it involves the nerves inside the hypothalamus. This hypothalamus is the same organ that appears to be so involved in the generation of sexual arousal (it is smallish, as you recall, and located near the base of our brains). There even appear to be job assignments for hunger in various regions. Destroy one area of the hypothalamus in a laboratory animal (that ventromedial region we discussed in the last chapter) and it won't stop eating, soon becoming obese. Destroy another area of the hypothalamus (the lateral hypothalamus) in a laboratory animal, and the creature won't start eating, soon suffering weight loss. Electrically stimulate an intact lateral hypothalamus and the animal begins eating again. Such experiments led scientists several years ago to the hypothesis that there are "I'm full" and "I'm hungry" signaling regions in this multi-talented area of the brain. At the time, it seemed that the only thing remaining for us to understand appetite generation was simply finding the various neural inputs and outputs to and from the hypothalamus. Sensors in the body might assay the energy requirements of the organism, make some interpretations and then relate their findings to this tiny region, creating a response.

Fat chance that anything in appetite biology would turn out to be so straightforward. It was soon discovered that body mass could be easily maintained even in the face of greatly varied nutritional supplies. This meant that additional cellular and molecular regulatory pathways lay undiscovered. Moreover, it was shown that both physical activity and energy intake could independently affect the energy balance of the body. It was thus thought possible that signals involved in eating and signals involved in human activities such as exercise could interact with each other. This turns out to be correct, but the interaction is so complicated that, to this day, nobody really understands the relationship between physical activity, energy intake, and the creation of hunger. Like I said, we know that it involves nerves, and we know that the neurons in the hypothalamus are deeply embroiled in the process. But the complexity hints at other, more peripheral signals involved in appetite generation, and most of those are a complete mystery.

What kinds of signals could influence the afferent and efferent signals impinging upon and responding to neurons in the hypothalamus? There really do appear to be sensors in our bodies informing us of our need for fuel. Many of them exist outside the brain and in certain lower tissues such as the liver, and it is to the liver that we now turn to try to describe how some of this interaction occurs.

The liver

The liver, as you know, is that red-brown blob of tissue tucked underneath the right side of our diaphragm. It derives its color from the fact that so much blood flows through it. In fact, it has two blood supplies, one from an artery (carrying oxygenated blood) and one from a vein (carrying newly absorbed nutrients). This busy trafficking makes the organ an ideal place to assess the contents of the blood and tell any tissue who is listening what it finds. In other words, it's a great place to have a sensor.

What exactly is a sensor? A sensor has two functions in the body: one to assess a particular event, the other to communicate something about the event to a relevant place, usually with some kind of consequence. In many ways, the entire *Divine Comedy* is about sensors, in this case, medieval moral ones. Heaven makes an assessment based on someone's behavior, the assessment is communicated to various custodians working in the afterlife and, consequently, the person is consigned to a specific place after death. To illustrate the function of sensors, consider one fairly pointed story about a person Dante encountered in the Gluttony Level, a person you might never expect to be there – in fact, a Pope.

As discussed previously, Dante is being guided through the Sixth Cornice by an actual friend of his named Forese, who had died five years before *The Divine Comedy* came into existence. Enshaded in Purgatory, Forese pointed out to Dante the various identities of the emaciated inmates. Near the end of their journey together, Forese identified one man possessing a face "more sunken-in than any other". The face belonged to none other than Simon De Brie of Tours, the French Pope Martin IV, who reigned from 1281 to 1285.

The heavenly sensors found him to be gluttonous because of a peculiar love of a certain food, which Dante described as "Bolsena's eels and the Vernaccia wine". These eels, found in Lake Bolsena near Viterbo, are still prized today as delicacies in the twenty-first century. In Dante's time, the eels were prepared by dumping them whole into a vat of Vernaccia, a sweet white wine made in the mountains near Genoa. In this way they were pickled alive; after a certain amount of time, the eels were removed from the vats and roasted. Martin IV could not eat enough of these, apparently, and eventually died from an attack brought on by his "overindulgence". He was immediately escorted to Purgatory by the Angel Boatman, and assigned to the Gluttony Level. In this mini-narrative we see the functions of a sensor: an assessment, a communication, and a consequence that brings about a result. Not incidentally, we also see the attitude some Italians had towards a "foreign" spiritual leader. Martin IV was not Italian.

In the liver, as in Purgatory, sensors also provide an assessment, a communication and a consequence. As we discussed, there is a lot of blood flowing through the liver, and the liver is very good at discovering its contents and signaling what it finds. Quite literally, there are neurons in the liver that assess the contents of the blood, communicate what they observe to the brain, eventually resulting in a change in eating. That's one of the ways the body can interact with the brain in the overall assessment of fuel consumption and expenditure. But, of all things, why assay the blood? What do the liver sensors look for as the red stuff courses through its tissues?

The answer to that question has to do with the way we digest food. Consider Martin IV, just before he dies, digesting his pickled and roasted Lake Bolsena delicacy. Like the rest of us, Martin IV has no way of directly extracting the nutrients in complex foodstuffs such as eels. Instead, he is forced to crush the eel down to its constituent parts, then internally slurp up the nutrients after the destruction. The Pope possesses levels of destructive firepower, just as we do. The teeth and saliva start the process with coarse crushing, and a combination of enzymes and physical action in the stomach turns most of the crushed slippery fish into an even more slippery slurry. After a few more manipulations, what's left arrives at the small intestine. It is there that the individual nutrients are taken up by certain cells, and are eventually tossed into our bloodstream. Because the nutrients go directly from the small intestine into our bloodstream, a sensor that can detect foodstuffs in the blood is a very handy thing to have.

And just exactly what do the liver sensors detect? Historically, there has been a fair amount of controversy about the interaction between meals and cells and the molecules involved in creating a hunger at this level. Some neurons in the liver, collectively known as hepatic afferent neurons, appear to act as sensors for sugars in the blood. They assess the amount of the sweet stuff (glucose really) by monitoring their own ability to eat it. If Martin IV skipped breakfast, for example, he would experience a drop in his normal amount of blood glucose. This less-sweet blood would flow into the liver, where it would encounter the sensors. A dip in blood-sugar concentration means less glucose for the neuron sensors to eat, and the sensors would fire off a signal to the Pope's brain, informing him of the loss. In a complex series of responses, the Pope would eventually feel hungry and go looking for his eels.

The liver has other sensors that work with blood contents besides sugars, and these have the ability to affect hunger cues as well. There appear to be sensors that can detect the presence of specific amino acids, for example, those building blocks of proteins we discussed in the last chapter. Many of these sensors, while shown to exist, have not been isolated. How those signals reach the brain and contribute to the cause of hunger is an even deeper mystery.

The liver informs the brain of the body's nutritional status in many ways, and this information somehow contributes to the feelings of hunger (there are so many functions the organ mediates that one could write an entire book just describing the major players). So what is the point in mentioning the liver at all? The liver is physically separated from the brain by about thirty-eight centimeters or fifteen inches in most people, an infinite distance when viewed from the level of the molecule. In order for the liver to interact with the brain, it is forced to hurl its information quite a way. As we have seen, some of this long-distance communication occurs via a system of interacting neurons. A nerve cell literally fires up a signal in response to a perceived metabolic stimulus, creating a chain of excitement that eventually reaches the brain.

However, there are other kinds of tissues besides the liver involved in the process of informing the brain of the body's nutritional status. They also appear to use long-distance communication devices to get their message to the brain. Curiously, these devices are not neural-based at all. As an illustration of this phenomenon, we will consider one other tissue that is vitally involved in the process of creating feelings of hunger – fat.

The role of fat

One talent we did not discuss concerning the liver concerns energy storage. The organ actually keeps a short supply of sugar as a mini-backup for people like Martin IV when they skip breakfast. But it's only a very short supply, and if Martin IV ever got lost in a desert, he would need a lot more than what's in the liver to help him survive. The solution? The body can create that nemesis of Western Civilization, good old pudgy, soft, water-logged, I'm-gonna-have-a-heart-attack, fat. In the parlance of boring scientists, fat is usually called triglyceride.

Though we know that if we exercise and eat right we can shed a great deal of our triglycerides, a question many people ask is, "why does it have to exist in the first place?" Besides making us feel self-conscious when we look in the mirror, what does fat really do? Though much maligned, fat is really a very necessary part of a mammal's survival strategy, and arose out of a need to solve an important problem. All organisms require energy to be continuously available, and that's a problem because the supply of energy in the environment is notoriously unstable. It would be nice to be able to store excess energy in some kind of biological battery, and then save it for a rainy day. And that is exactly what most animals do. The battery most mammals came up with was good old fat, stuffed not in metal casings but in bloated cells called adipocytes. If the creature needed energy and no food was available, it might just dip into its energy supply tucked inside these adipocytes. When excess food became available, it might then restock the leftovers by turning them into triglycerides. We humans can luxuriate in the fact that we have, for the most part, stabilized our supplies of energy. And that's why many of us not so luxuriously worry about becoming obese.

If mammals really can create biological batteries to save for a possible famine, certain types of communication become very important to the animal. It needs: (1) a mechanism for assessing the information at the level of energy stores; (2) a mechanism for communicating this assessment to important regulatory regions in the body (like the brain); and (3) a mechanism by which the brain regulatory regions might act to influence the ratio of energy intake versus energy expenditure. The feelings of hunger (and its opposite satiety) come somewhere between (2) and (3).

Do mechanisms that keep the brain informed of its back-up supplies exist? Might such mechanisms be deeply involved in the decision to create hunger-related feelings? The answer to both questions appears to be yes. In many cases, the processes appear to be mediated not by neurons but by hormones, many only recently discovered. These hormones are created by genes, and as more genetic sequences are isolated a clearer picture is beginning to emerge. In many cases, these gene products target those regions in the hypothalamus we discussed previously. To understand how this biology relates to the brain and to feelings of hunger, we will talk about six newly discovered biochemicals and, most importantly, the genes that make them.

A molecular view

It is admittedly a hard task to look for genes governing hunger behavior. One reason is that, from the point of view of genetics, there are several ways to look at the generation of appetite. One might say that hunger occurs because some inhibitor of hunger feelings has been destroyed, perhaps due to the absence of fuel or back-up batteries. One could also say that hunger occurs because some new gene product has arisen, either from the brain directly, or acting upon the brain from an outside source. Considering the complexity of hormones acting on the whole situation, both models might occur. Indeed, such a synthetic picture may be the most accurate way to look at how hormones work to generate and respond to feelings of hunger.

For reasons of space, we are forced to confine our molecular discussion below to genes that are known to interact with cells inside that hyper-busy hypothalamus. Ultimately we will be talking about molecules affecting not only the generation of hunger feelings, but also feelings of satiation (after all, destruction of such satiation molecules could just as easily unleash an appetite). The questions posed are simple ones. How does the hypothalamus do it all? What are the molecular signals that help inform the hypothalamus about our food requirements? How do they communicate with the rest of the brain?

As we'll see, it is a lot easier to ask these questions than to answer them. To begin our journey into the molecular world, we first have to address a mechanism common to all six molecules. Each participates in a series of startlingly complex series of feedback loops, some only recently discovered. A feedback loop is actually a scientific term, and we will use an illustration to help us understand it. Specifically, we will talk about something Dante and his friends encountered along the road of the Sixth Cornice, that fragrant and somewhat mysterious first tree mentioned at the start of this chapter. Then we will return to our six molecules and discuss in detail their role in the generation of appetite.

Punishment in a plant

As was described previously, Dante and Virgil and Statius are walking along the ledge of the Gluttony Level when they encounter a very interesting tree, growing right in the middle of the road. Dante describes it this way:

But soon in mid-road, there appeared a tree
Laden with fragrant and delicious fruit
and at that sight the talk stopped instantly
As fir trees taper up from limb to limb,
so this tree tapered down; so shaped, I think,
that it should be impossible to climb

This sounds like an odd piece of botany, with the last two sentences appearing especially interesting. Older commentaries have suggested that the tree might actually be growing upside down, and is even illustrated that way in some old manuscripts.

What is not cryptic is the purpose of the tree. As the three poets gaze upwards, they hear a voice from amid the tasty fruits. The voice cries a warning, telling all sinners to stay away from sampling the fruits (reminiscent of The Tree of Knowledge in the biblical Garden of Eden, a root of which the travelers encounter later). After the warning, the voice describes great examples of moderation of appetite, mostly Biblical in nature. The tree is obviously some kind of information distribution center, with words appearing to address the behavior of the various gluttonous souls of the Sixth Cornice. Indeed, the emaciated souls of the Sixth Cornice hear these words as they run past the tree, completing their endless circuit. Dante recognizes yet another old poet friend in the pack, who stops to explain the interaction of the souls with the tree.

Hunger and thirst that nothing can assuage
grow in us from the fragrance of the fruit
and of the spray upon the foliage
And not once only as we round this place
do we endure renewal of our pain.
Did I say "pain"? I should say "gift of grace".

It seems that every time the emaciated gluttonous shades of this level complete a circuit and come back to the tree, their hunger and thirst are renewed, their pain sharpened, their desire for the fruit ever intensified.

That, of course, is their punishment, and Dante's friend is describing what scientists would call a feedback loop, one meant ultimately to inform the souls of their unrighteousness. The more times they encounter the tree, the hungrier they get, and the more they desire the forbidden fruit on its branches. The more hunger they feel, the more they want the fruit, and so there is this never-ending spiral of misery. Such repetitive experience of information is the heart and soul of a feedback loop. The souls appear to chafe at the lesson, allowing the almost bitter allusion to pain being the "gift of grace".

The metaphor of a tree involved in a feedback loop is used here to illustrate some of the ways specific signals work with the hypothalamus to influence feelings of hunger. There are, however, as many positive feedback interactions as negative ones, and some very exciting recent research has begun to describe these loops at the tissue and gene level. While we could spend the entire book just cataloguing these results, space dictates that we stick to a few of the major players involved.

Looking first at obesity

When researchers started listening to the major themes of hunger, they concentrated their efforts on factors that might influence food intake. An obvious strategy was to examine easily assessed abnormal situations – like mutations – discover what was wrong, and then start building the story. In terms of food intake, the most obvious abnormality is either complete starvation or uncontrolled obesity. Explorations of the latter turned out to be especially fruitful, and completely frustrating. The reason for the anguish is that scientists got two tantalizing clues about the genetics of obesity, but could not explain them for almost half a century. Here's the first: in 1950, a genetic defect was discovered in laboratory rodents that caused them to become quite obese. These mice did not know how to stop eating and could easily weigh two to three times more than their unaffected litter mates. The trait – and the mouse – was called ob. Here's the second: if an ob mouse was surgically joined to a normal animal (a process called parabolic surgery), the abnormal ob mouse started eating less and eventually lost weight. This led certain researchers to hypothesize the existence of a "satiety" hormone in normal mice, one that directly controls appetite. The ob mouse lacks this hormone, and the above experiment was explained by postulating that the normal mouse resupplies the substance when the two are surgically connected.

It took until the mid-1990s for the molecular explanation to reveal itself. The hunch from the surgery was the correct one: there truly was a circulating hormone, which was eventually isolated and called leptin (see Figure 9). Once that molecule was isolated, a whole flurry of research results followed. It was found that leptins worked in cells by first binding to some kind of receptor. Consistently, many of these receptors were found on the neurons of the hypothalamus. If that sounds like the equipment for some kind of feedback system, you are in agreement with many scientists. Two overall models for this interesting gene product in our feelings of hunger have been postulated:

Role 1. Response to starvation. In the fed state, leptin is normally pumped directly into the circulatory system from those adipocytes (fat cells) described previously; that's how the hormone can reach the hypothalamus. If leptin levels fall off, however, a specific set of adaptive responses occurs in laboratory mammals. The animals acquire a diminished reproductive capacity and levels of certain other hormones – like thyroid hormone – go down as well. This has the net effect of conserving energy, a condition very much needed in order to survive certain stressful environmental situations (you don't have to conserve energy if there is a proper and abundant food supply, but you definitely need to if there isn't). Because starvation is usually more of an immediate threat than obesity, some researchers believe that leptin's primary role is its response to a lack of food. That means it is deeply – perhaps inadvertently – involved in the overall creation of appetite.

Role 2. Signaling satiety. Since the absence of the hormone produces obesity, it appeared obvious to many researchers that leptin controls appetite. A current model suggests that leptin levels rise with increasing obesity. As more leptin is dumped into the circulatory system, more of it becomes available for engaging those receptors found in the hypothalamus. Given enough occupancy, the presence of leptin on a critical number of receptors may be the source of the "I'm full" signal, and certain autonomic functions (remember those?) may begin to kick in. Such functions could include increased heat production, alteration of hormones such as insulin, and sequestering a body's energy stores away from fat. The processes all work together in an attempt to keep the animal from getting too fat. In the parlance of the research world, it is called "resisting obesity".

Whichever role leptin turns out to mediate, what is clear is that research into the little molecule is revealing an extraordinary series of interlocking feedback loops, all in molecular terms. If a laboratory animal gets too full,

The molecules of appetite contr

*Many biochemicals are involved in the regulation of human appetite.
below is how one such molecule, called leptin, is thought to work.*

LEPTIN AND OTHER MOLECULES

Leptin appears to control routes that (1) lead to stimulation of higher centers in the brain ar
control other tissues of the body. Such double-duty is performed by targeting regions in the
hypothalamus, areas which we have analogized to "concert halls" in the text. Four important
are the arcuate nucleus, lateral hypothalamus, the dorsomedial hypothalamus and the paraven
nucleus. How leptin interacts with these complex sounding regions is shown below.

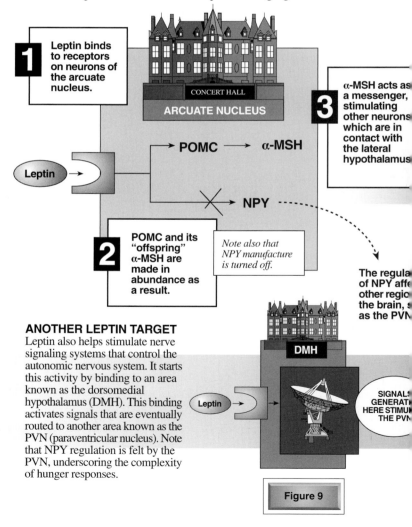

1 Leptin binds to receptors on neurons of the arcuate nucleus.

CONCERT HALL
ARCUATE NUCLEUS

3 α-MSH acts as a messenger, stimulating other neurons which are in contact with the lateral hypothalamus

Leptin

POMC → α-MSH

NPY

2 POMC and its "offspring" α-MSH are made in abundance as a result.

Note also that NPY manufacture is turned off.

The regula
of NPY affe
other regio
the brain, s
as the PVN

DMH

Leptin

SIGNAL!
GENERAT
HERE STIMU
THE PVN

ANOTHER LEPTIN TARGET

Leptin also helps stimulate nerve signaling systems that control the autonomic nervous system. It starts this activity by binding to an area known as the dorsomedial hypothalamus (DMH). This binding activates signals that are eventually routed to another area known as the PVN (paraventricular nucleus). Note that NPY regulation is felt by the PVN, underscoring the complexity of hunger responses.

Figure 9

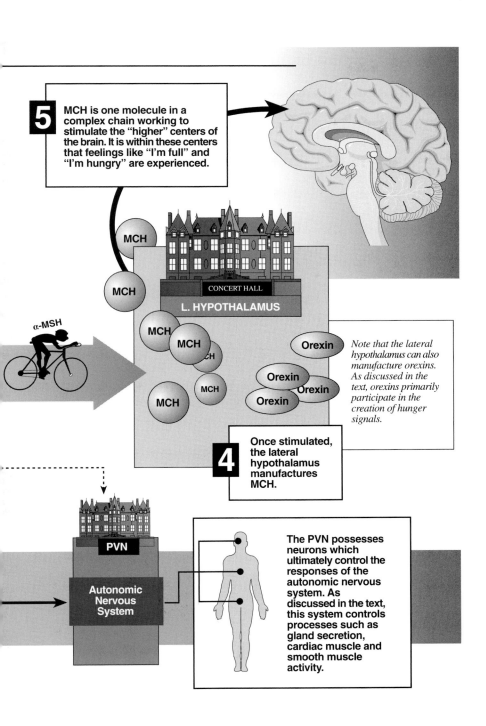

5 MCH is one molecule in a complex chain working to stimulate the "higher" centers of the brain. It is within these centers that feelings like "I'm full" and "I'm hungry" are experienced.

MCH

MCH

α-MSH

MCH

MCH

MCH

CONCERT HALL

L. HYPOTHALAMUS

MCH

MCH

CH

MCH

MCH

Orexin

Orexin

Orexin

Orexin

Note that the lateral hypothalamus can also manufacture orexins. As discussed in the text, orexins primarily participate in the creation of hunger signals.

4 Once stimulated, the lateral hypothalamus manufactures MCH.

PVN

Autonomic Nervous System

The PVN possesses neurons which ultimately control the responses of the autonomic nervous system. As discussed in the text, this system controls processes such as gland secretion, cardiac muscle and smooth muscle activity.

leptins come along and soon the animal quits eating. After that, the leptin levels fall, the animal uses up some of its available energy stores, and begins to lose weight. Hunger is eventually felt, which means the animal eats. In response to the fed state, leptins turn back on, the animal quits eating and the loop starts over again, round and round like inmates circling the forbidden tree on the Sixth Cornice.

But what about humans?

As we stated at the beginning, most of this work on feedback loops has been done with laboratory animals, primarily rats and mice. Does any of this have relevance to humans? After all, we humans can get obese, respond to starvation and possess thyroid and insulin hormones. We even have a hypothalamus. Do leptins play an important role in the management of our hunger too?

The answer, as you might expect, is somewhat confusing. Even though we have some very similar molecular equipment, we humans have to solve certain special problems that rats and mice don't. And note the use of the word "similar" as opposed to "identical". Genetically, of course, we are only somewhat related to rats or mice. These differences have not stopped investigators from trying to find important answers, of course, and research into leptins and humans has proceeded with enthusiasm.

And what has been found? In humans, you cannot increase leptin levels by eating food. That's not true of the familiar hormone insulin, whose levels are directly related to the occurrence of eating, rising and falling like a soufflé. Nonetheless, there is research suggesting that insulin might regulate the manufacture of leptin in certain kinds of fat cells. We also know that obese people tend to have elevated levels of leptin in their bodies relative to slim people. It is almost as if the body were creating signals attempting to slow the progress of the weight gain.

Curiously, there is a genetic disorder that involves a mutation found in the leptin gene. People who have this mutation have little functional leptin in their bloodstream. What do these people look like? They are huge, exhibiting what doctors like to call morbid obesity. Such an experiment of nature is just like the ob animals, who also genetically lack the leptin gene. This, of course, has tempted investigators to say that the roles of leptin between the laboratory animals and humans may not be so different after all. There may be just a few more variables to consider in us more complex creatures. In my view, the jury is still out concerning the role of this fascinating gene in our appetites, whether involved in starvation or satiation.

There is less controversy surrounding the targets that leptins seek when they leave a fat cell and enter the bloodstream. Leptins are looking for receptors in the brain, trying especially to communicate with (no surprise here) the hypothalamus. Once leptins arrive at their brainy targets in the hypothalamus, a complex series of mechanisms is triggered. These mechanisms create events which lead to changes in eating behavior. To understand exactly what happens, we have to describe a few processes that probably occur simultaneously. Consider it to be something like a molecular motet, with a number of voices in the hypothalamus going on all at once as soon as the leptin announces its presence to the organ. What I'll do is list a few of the main "singers" inside certain nerve cells, and then discuss what they are saying in response to leptin. The first three singers are found in one area of the hypothalamus, a tiny region known as the arcuate nucleus. The last three are found in a general area known as the lateral hypothalamus. (Leptin also binds to an area in the hypothalamus known as the dorsomedial hypothalamus, where it regulates part of that autonomic nervous system we discussed previously. For a detailed explanation of this anatomy and the gene products that follow, see Figure 9.)

Into the arcuate nucleus

The first singer is called hypothalamic NPY (neuropeptide Y). This very interesting molecule probably plays a direct role in the mediation of hunger. When scientists repeatedly injected NPY directly into the brains of laboratory animals, it was almost as if they had recreated the ob mouse. The animals got very, very fat. Since leptin's job in the fed state is to keep animals from becoming obese, one might expect that leptin normally shuts off NPY in certain regions of the brain. That is exactly what researchers have found. When leptin arrives at the doorstep of certain brain neurons in the arcuate nucleus, it immediately goes to work turning off the "voice" of NPY, which helps to regulate appetite.

POMC and MSH

The second and third singers actually form a duet, a father-and-son routine, capable of decreasing appetite. The father is called POMC (which stands for proopiomelanocortin) and the son is called α-MSH (α-melanocyte-stimulating hormone). The son is the talented one of the bunch. If he is around, he can communicate the "thanks, I'm full" message to regions of the hypothalamus that need to know. POMC is a molecule found in about one in three of all neurons within the arcuate nucleus of the hypothalamus. It is really a precursor molecule capable of producing the offspring α-MSH, hence the analogy.

What does this father-and-son duet have to do with appetite? It turns out that leptins can both arrive *at* and bind *to* neurons possessing POMC. This binding stimulates the neuron to make lots more of Dad. With lots more of Dad (POMC) around, the son is sure to follow, and pretty soon the cell is singing the duet, making both POMC and α-MSH. With the presence of the son in hand, the hypothalamus begins to understand that the body has had enough food, and soon the animal stops eating. As you can see, there are both negative and positive consequences of leptin's binding to the arcuate nucleus.

The lateral hypothalamus and higher centers

So what happens next? Obviously, there are lots of biochemicals creating hunger themes at the molecular level, all built towards generating and regulating our appetite. However, the arcuate nucleus is not the only concert hall where hunger-singers perform. A region known as the lateral hypothalamus is also involved. What is the connection between the two concert halls, the arcuate nucleus and the lateral hypothalamus? The connection occurs because α-MSH, that talented son of POMC, binds to neurons which eventually stimulate cells within the lateral hypothalamus. The son is really acting like a courier to the lateral hypothalamus, bringing it news of energy balance.

Why, though, discuss the message path to the *lateral* hypothalamus? We do this because the *lateral* hypothalamus acts like a newspaper's reviewer of popular arts. This region is "read" by many smart parts of the brain, those higher centers we have been discussing. And it is in these higher centers that our feelings of appetite are experienced. So the question is, how does that work? Are there biochemicals involved in this transfer of information from the hypothalamus to the rest of the brain? The answer turns out to be yes, at least in part, and the chemical anatomy constitutes our next group of molecular singers.

Believe it or not, the first member of this new group was found in a salmon. It was called MCH (short for melanin concentrating hormone). This odd biochemical actually helps regulate skin color in fish scales; however, it was also discovered that it exists in animals, and performs a vastly different function in these land creatures. It was discovered that if you inject MCH into specific areas of the brain in rats, you can stimulate their feeding. This means that, somehow, it is involved in appetite. However, this biochemical was not found in the same concert hall as the other three molecules mentioned. Rather, it is confined to some very interesting neurons in the lateral hypothalamus. Why do I say interesting? Because the MCH-laden neurons are connected to "higher" regions in the prefrontal cortex, the area where the emotion of hunger might be experienced. All of a sudden, researchers found themselves working with a molecule made in the hypothalamus, but capable of interacting with the smarter regions of the brain.

A real gene for appetite

The last two molecular singers to be discussed in this section, a twin act, may represent the most direct line to appetite generation we have yet discovered. So far, by discussing the roles of leptins we have been talking about appetite in an almost indirect fashion. A number of researchers think that leptin's primary role has to do with a creature's response to a lack of food, as mentioned previously. Since most mammals in the wild do not have to deal with obesity (nature seldom provides years-long excesses of foodstuffs), emphasis on the more common experience makes a great deal of evolutionary sense. But recently, a series of gene products and their receptors have been isolated that appear to work in a much more direct fashion. Some of them also appear to work in the lateral hypothalamus, and work to create signals to the higher regions as well. To illustrate their role in energy balance, I would like to turn away from complex biochemistry and use as an illustration a description found in *The Purgatorio*. It has to do with Greek mythology in association with the physical appearance of those souls in the Sixth Cornice.

Dante and his friends have just finished encountering their first tree, with the tortuous voice emanating the moderation homily. As mentioned, Dante then hears a frightening chorus of inmates behind him, singing a hymn that appears to be more like a cry of pain. And they look horrible. Consider how Dante describes them:

The sockets of their eyes were caves agape;
Their faces death-pale, and their skin so wasted
That nothing but the gnarled bones gave it shape
I doubt even Erysichthon's skin,
Even when he most feared that he would starve,
Had drawn so tight to bone, or worn so thin.

The Greek mythology reference in this stanza is to Erysichthon, giving a terrifying poignancy to the description of the inmates. Erysichthon was an arrogant man in Greek legend who had the gall to mock the goddess Ceres (goddess of agriculture). He did so by taking an axe to one of the oaks in her sacred grove and felling it. Ceres was so furious that, as punishment, she gave Erysichthon an unquenchable hunger. And she did make it *insatiable*, no leptins allowed. Erysichthon soon ate up everything he owned, put his daughter up for sale, and used that money to buy more food. When that did not satisfy him, he chopped off his own limbs and ate them as well, consuming himself in the process. Dante uses this as a great metaphor for the punishment of the gluttonous, who devour their own souls just as Erysichthon devoured his own body.

The interesting lesson in this story has to do with insatiability, a process so compelling that not even Erysichthon's own flesh was immune. Though Greek mythologists would have absolutely no idea of this, the biology they describe was not far from the mark. Quite recently, specific gene products and receptors have been isolated that appear to create directly the need to consume and consume and consume. Unlike leptins, these signals do not exist to alter feelings of being full; these molecules help create the "I am hungry" sensations. The role these gene products play in the creation of appetite is important enough that it is worth describing how they were discovered.

How orexins were isolated

Talking about the history of appetite generators is really a collision between two styles of science, generalists and specialists. Energy balance is really a problem involving many kinds of tissues all over the body. Yet the tiny molecular interactions of which they are composed are so complex, one can generate an entire scientific career around understanding just a single gene. There is a danger in that, because it is easy to lose the big picture. In the real world of living organisms, there are batteries of cells making large numbers of interactive, integrative biochemicals. That's why in our discussion of appetite I have continually emphasized the complex and disparate nature of the processes involved in maintaining an energy supply.

Some molecular researchers have chosen to stay in the big picture as much as possible. Many in their careers have isolated entire classes of genes whose functions are completely unknown, the division into classes based solely on external structure. A case in point is the isolation of genes that encode a collection of receptor proteins (like receptors involved in making appetites), without knowing anything about what they do. Termed orphan receptors, these collections of molecules are found on the surface of certain cells. Scientists isolate these molecules, throw them into experimental cells as if the cells were packing crates, and put them all away in the freezer. They then may leave it to the specialists to go after specific functions.

Why are they called orphans? As we mentioned previously, a receptor's job description is to bind to other externally derived molecules (hence the term "receptor"); these external molecules are usually called ligands. Ligands fit into receptors like keys into locks, as also mentioned previously, and the cell that gets its receptors occupied often acquires a new function. In molecular biological research, there are ways to isolate entire banks of receptors, even if the ligand is completely and unhappily unknown. That's what an orphan receptor is, a receptor whose structure is known but whose ligand is a complete mystery.

What this has to do with appetite

The way the genes involved in the creation of appetite were isolated is really a story of collaboration between generalists, who enjoy isolating molecules of unknown function, and specialists working in the practical world of appetite biology. A group of generalist researchers in Japan isolated a collection of fifty of these orphan receptors from rat brains. Each receptor gene was placed into its own cell and allowed to express its encoded protein on the surface of that cell. The cells were then grown in culture (see Figure 10).

Of course, even a generalist would not let the research stop at the level of incubation. Finding functions for the orphans is paramount to the research effort. But where does a generalist start? Obviously, a receptor's function cannot be fully explained without knowing the ligand to which it binds. The question was, of course, what ligand? At this point so-called

Isolating hunger genes

Using various extracts from rat brains, certain genes that mediate appetite creation in mammals were isolated. Here's how it was done.

Techniques have been developed allowing researchers to isolate genes based on structure rather than on function. This means that the biological role of individual sequences isolated in such a fashion is unknown. In these experiments, genes were isolated that encoded previously undescribed receptor proteins. Such receptors are often called "orphans" indicating that neither the function nor the ligand (the molecule that binds to the receptor) has been discovered.

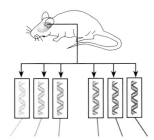

1 **PLACING THE ORPHANS IN CELLS.** The genes encoding the orphan receptors were then placed in the nuclei of specialized cultured cells. Once there, the genes were transcribed and translated. Eventually, the protein receptors were placed by the cell onto its surface.

2 **ASSESSING RECEPTOR ACTIVITY.** The cells were situated in such a fashion that their receptor activity could be monitored. Culture conditions were maintained so that if the receptor became occupied, the binding could be detected.

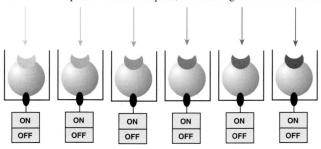

Figure 10

3 **ISOLATING BRAIN PROTEIN**. Rat brains were ground up and protein fractions isolated. Why do that next? The idea was that if the orphan receptors came from the brain, protein ligands binding to them might also come from the brain. Grinding up the entire tissue would ensure their isolation.

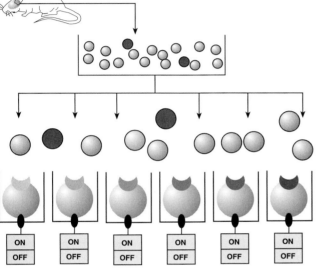

4 **ADDING PROTEINS TO THE ORPHANS.** Fractions of the protein extracts were added to the cells containing the orphan receptors. If the proper ligands were in the mixture, they should bind to a specific receptor. Using those specific receptor-occupancy monitoring devices, this binding should be detected.

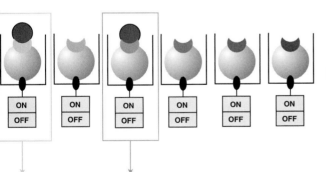

5 **LIGAND BINDING.** Proteins existed within the mixture that did indeed bind to the orphan receptors on the cell. These receptors were identified and the ligands were untethered and purified.

TESTING FOR FUNCTION. Using information derived from the ligands themselves, the genes that encoded these proteins were eventually isolated. Further tests showed that these genes were expressed in regions of the brain known to be responsible for hunger. The proteins encoded by the genes could induce hunger if applied to those very regions. Using these 5 steps, genes involved in the creation of appetite were isolated. They are now called orexins.

fishing expeditions take place, something at which generalist science types are really quite good. In order to isolate the binding ligand, the generalists performed a series of experiments, ones that could be divided into four overall parts. Here's what happened:

1. The cells carrying an individual orphan receptor type on their surfaces were grown in culture.
2. Measurement devices were set up to detect whether something bound to the orphan receptor. Without going into details, if a ligand binds to a receptor using certain cells grown under specific culture conditions, there are machines that can detect that binding. It is simply a way of identifying the "I gotcha" moment, the fish on the line, essential in work like this.
3. Crude extracts of rat brain protein were then made. The idea here was simple: if these orphan receptors do come from the brain, it is logical to look first for the ligand there as well. The researchers took two hundred grams of rat brain, put it into an osterizer (no kidding), and isolated the protein fraction from the pulverized goo. Having a test tube full of partially purified proteins containing ligands, even if you didn't know which one was the molecule of interest, was a very useful thing to have for the next step.
4. Representative protein extracts were then literally poured over the cells in culture and allowed to incubate. Each cell line in each culture carried a different orphan receptor. The hope was that if the ligand were in the extract, it might possibly bind to the orphan receptor sitting on its surface. This binding would be detected by the machine. If there were no ligand in the extract, no such binding would occur, and the machine would say so.

The main course

Following this recipe, the generalist researchers hit pay dirt. They actually found several proteins that tightly bind to the orphan receptors sitting on the cells. The next set of steps were then easy. After hosing the cultures down, the researchers coaxed the receptors to let go of their little protein ligands, which were then isolated and put into test tubes. Since genes encode proteins, if you have the protein sequences in hand, it doesn't take very long to isolate the gene. I'm skipping a bunch of steps here, but that is exactly what happened. The researchers used the sequence of subunits in the protein (amino acids) to eventually isolate the genes that encode them.

But now what? At first glance, finding the ligand doesn't seem to be much progress; you have orphan receptors and now you have orphan ligands. The most useful thing a generalist could do would be to ask this question: now that I have the gene for the ligand, and it binds to a receptor in the brain, where might the ligand be expressed normally? Perhaps in the brain as well? If you could actually pinpoint exactly where this gene was made, you might begin to get a handle on its function.

That, of course, is exactly what the researchers did, and they started by looking inside discrete regions of the rat brain. There are techniques available that can show a genetic engineer just exactly where a specific gene is being expressed. The researchers discovered that these genes were expressed primarily in the lateral hypothalamus, that familiar arts reviewer region of the brain we discussed earlier. That's a very exciting location in the brain to have such specificity of expression. Based on the functions of the region, all of a sudden the generalists had a set of testable hypotheses. One of them, of course, was appetite control.

Enter the drug companies

This is the point in the story where the generalists come in contact with appetite specialists. The Japanese researchers were already collaborating with a group of scientists at SmithKline Beecham; the marvelously specific expression profile of these new ligands gave everybody the potential to become appetite specialists.

Tucked inside this collaboration, generalist and specialist researchers both hypothesized that the new ligands helped mediate appetite. Since the expression was in the lateral hypothalamus, the research could be refined to speculation about the origin of food lust. The collaborative groups did two very simple experiments to ask questions about appetite:

1. Regulation. If the ligands mediate appetite, then gene expression of the ligands should be elevated in the lateral hypothalamus when the rat is starved, and should drop when the rat is satiated. That is exactly what was found. When rats were starved, their brains started pumping out large quantities of ligand. When the rats were satiated, the expression of the ligand went away.
2. Overeating. If the ligands mediate hunger, then injecting the ligand

directly into the rat's lateral hypothalamus should increase its appetite. As a result, the amount of food the animal eats per unit time should increase. This experiment was done, and once again the researchers hit the jackpot. The injected rodents ate three to six times as much as controls in a given time period.

The summary of the data pointed in a single direction. Here was a gene whose expression was regulated in a known appetite-mediating area of the brain. This modulation was specific: the gene was activated in that region when a rat was starving and down-regulated when the rat was not. The direct injection experiments provided the functional framework, demonstrating the same appetite-mediating pattern. With the success of all this work, and many other confirmatory controls, the researchers decided to name their ligands orexins, after the Greek word for appetite (*anorexia*, as you know, means loss of or without appetite). They had indeed found a ligand-and-receptor pair that was deeply involved in food intake. By the time all the experimental dust settled, they had actually found two of them.

I've spent a long time describing the events leading to this discovery for a simple reason: it represents a brand new perspective on the generation of hunger. As we have discussed previously, the presence of leptin gives us an indirect way of looking at the generation of appetite: the absence of a *repressor*. Orexins seem to work in the opposite direction, giving us the need to integrate the presence of a *stimulator* into the overall picture. We thus appear to have both repressing mechanisms and stimulating mechanisms in the generation of our hunger.

No unifying hypothesis

Taking stock of all these tissues, from brains to fats, and all these molecules, from leptins to orexins, is there any overarching idea that unifies them and gives us a perspective on how we experience hunger? Can we make a cogent model that puts the hypothalamus, liver sensors, POMC, leptins, and orexins into a single set of processes? The answer to both questions is "no", at least not at present. Even if we spent a thousand pages detailing every known piece of scientific data uncovered about appetite, we would only create a dim outline in the mysterious world of hunger generation. One researcher has gone on record as saying the following depressing statement:

... hunger and satiety are not the whole story, and may not even have a privileged position in the story ... Perhaps we should celebrate the fact that we do not know why animals or people eat the way they do. At least we can boast of knowing less than the experts who hold forth at cocktail parties!

(Herman, 1996 p. 17)

All we can really say is that the creation of hunger is an aggregate effort of many players. We know that tissues in the body work with tissues in the brain to create energy balance, and that the communication involves both neurons and hormones. We also know we often feel hungry when the brain tosses up what it knows about our energy situation to higher cortical centers. But, like a chorus in the distance, we can detect that the music is there, but have no real idea what the voices are singing about.

More about minds and brains

Part of the reason appetite feelings are so mysterious is that so many of the variables are undefined. There is a large part of this discourse, for good or ill, that is involved directly in the no-man's land of minds and brains, an area where few biochemists want to tread. Historically, they are not alone. As we discussed in the last chapter, the people of Dante's century would have considered the notions of mind and brain to be as foreign as the sin of gluttony familiar. Borrowing from the Greeks, the medieval mind thought about human creation in terms of a soul within a body, not a chorus of neurons creating a brain. In the Sixth Cornice, this description led Dante to ponder a contradiction: he wondered aloud how *shades* could appear to be starving, when in fact they were made of airy bodies and did not require food. The answer to that question led to Statius' long discourse on the nature of human beings, providing hints to the medieval perspective of human anatomy, both spiritual and physical.

The idea of a soul within a body began to change as the years went by, however. More than two centuries after Dante's death, the restlessness of intellectual life in Europe was growing stronger, and that meant challenging many long-held assumptions. Copernicus and Galileo were changing the way we looked at the universe. Newton eventually uncovered physical laws with such insight that the basic precepts remain accurate to this day. The nature of humanity, both physical and spiritual, was not immune to such vigorous challenge and redirection.

Into this intellectual stew was born Rene Descartes, who gave us our next set of ingredients for understanding the history of mind and brain. In fact, he framed the question, then gave an answer that still gives many pause today. Essentially, he said that human beings were composed of two kinds of substances, a body and a mind. The body was quite divisible, more like a magnificent machine, but one that could certainly break down, even die. The mind, on the other hand, was not divisible. The mind instead was a conscious being, one that willed things, sensed things, imagined things. Because of its indivisibility, it could exist without being extended in three dimensions, and so might survive when the body died. As the human itself possessed both of these parts together, his notions became known as dualism, and his famous insight "I think, therefore I am" placed the essence of humanity on the mind-side of the equation.

It is very easy to see both the Greek and the medieval roots of such thinking. And it is also easy to see that the model created one giant problem for Descartes to solve. Obviously, the mind and brain must interact at some level. But how could something physical like our bodies be moved by something nonphysical, like our will? How could the physical smell of some precious fruit bring to us feelings of hunger in our – what – minds? The job was made tougher because Descartes had to integrate the ideas of a mechanistic biological universe with the teachings of the Church. Having watched the travails of Galileo in his attempt to do the same, he became naturally reticent to display his full thinking on the matter.

Not that he didn't try, of course, for the problem of interface was as much an intellectual challenge as it was a political highwire act. Descartes began a search for some area in the brain that could act something like a quiet cafe, whereby both mind and matter could sit down and talk to one another. He settled on a region now known as the pineal gland, which Descartes called the third eye. He suggested that this gland was the place where the soul entered and could interact with the body. In modern terms, we might be tempted to call this an interface. Descartes was wrong about the pineal gland, of course. It has functions more associated with sleep/wake cycles than with a watering hole where neurons can meet an attitude.

Descartes and another ingredient for consciousness

Regardless of the contradictions that lay before Descartes and eventually his posterity, placing human identity on the shoulders of human thought was quite an achievement. It still is in many ways: taking stock of the actions of the brain and infusing them with identity is a very twentieth century thing to do. We have wrapped some of his ideas in fancier terms, using words like consciousness, for example, in describing parts of the "I am" statement. And since the experience of hunger involves awareness at some level, it is appropriate to address it here.

It is always astonishing to me the number of memories that I can conjure when I smell a favorite food being cooked. Undoubtedly, neural regions holding long-term memories of distant meals are being stimulated by such smells. In order for those long-term associations to be brought into my awareness, however, something else has to happen first. That something else involves the immediate recognition of what is being cooked. Without an initial comprehension, I would never be able to summon the past memories that stimulate gustatorial nostalgia. In other words, some kind of initial awareness is needed before I can start reminiscing. Scientists call this initial awareness – the ability to absorb new input for a small period of time – short-term memory (or, perhaps more accurately, short-term buffers). It is the gateway that must be passed through before memories can be stored and retrieved, and, as such, is a very important part of the definition of consciousness. It is the next ingredient I would like to explore in our attempt to identify some of its most important components.

To get an idea of what a short-term buffer is, consider the following three examples.

You recall that the trio of poets has encountered two trees in the Sixth Cornice. A voice emanates from the first tree, giving a warning not to eat any fruit, a voice whose origin Dante becomes curious to find. Virgil ignores Dante's curiosity, and admonishes everyone to keep walking, lest they waste time. Dante turns from his inquiry to follow his two companions.

How does this describe a short-term memory buffer? It may seem almost too obvious to mention, but in order for Dante to obey Virgil's voice Dante had to have an area in his brain where Virgil's commands could be temporarily stored. In other words, there had to be some kind of transient buffer active in Dante's brain. If Dante did not possess this temporary storage area, Virgil could talk all he wanted, but Dante would pay attention only as long as speech was heard. With no temporary buffer to store the memory of the words, Dante might easily ignore the Roman's warning and start doing something else as soon as Virgil stopped talking. In other words, Dante had to possess a temporary storage buffer.

Here's another example of short-term buffer. Read the following partial sentence, a part of the poem describing when the poets encountered the second tree in their journey, taken from Eden's Tree of Life.

We turned a corner, and there came in sight,
Not far ahead, a second tree, its boughs
Laden with fruit, its foliage bursting bright

While reading that last sentence, you probably remembered that the boughs were laden with fruit, even though it also contained foliage bursting bright. You might even remember that the poets had to turn a corner to discover the tree with all its bright leaves. These memories occurred even though you are currently reading this sentence, and not the lines from *The Purgatorio.* That ability to remember is another example of a short-term buffer. You could recall the fruit, even when you were reading about foliage, because a short-term buffer temporarily held in memory the words around "fruit" for you, allowing you to understand the entire sentence. You could keep the first part of the stanza in mind even though you were reading the last part, or even this paragraph, because of these handy buffers. Thus, not only are there short-term buffers for remembering voices, but also for remembering sentences.

Here's one last example. As the emaciated prisoners of the Sixth Cornice encounter both trees in their circular jogging, they become dismayed. Why? There is olfactorial input (the fruit is fragrant) and visual input (the fruit looks tasty), and then there is verbal input telling the inmates that they cannot eat. The dismay occurs the instant they hear the harsh words because of the presence of at least two other short-term buffers: one for the memory of the smell of the fruit and one for the memory of its sight. These inputs are compared to the verbal commands the inmates just encountered, and emotional reaction occurs as a result of the evaluation. The interesting aspect in this example is that the memories of all these events were held simultaneously; the olfactory memory was not forgotten simply because they also saw the fruit, nor was either memory lost the instant they heard the voice. Rather, the presence of the various inputs occurring in parallel allowed the disappointment to take place.

Taken together, these cognitive strings-on-a-finger have two very important characteristics. The first is that specialized buffers exist for many (some researchers would say all) our sensory inputs. Understanding language, reading sentences, smelling and seeing fruit all require specific sensory processing. Yet they can be integrated into the experiences of the Sixth Cornice because of the existence of distinct and attached short-term memory buffers. The second is that these buffers can hold their pieces of information simultaneously. They function *independently* of one another, without one sensory memory crowding out another simply because the input it processes is different. It appears that all active thinking processes use these temporary buffers, doing so when specific types of inputs have to be retained in order to make sense of the world.

What this has to do with consciousness

What is the relevance of short-term buffers to consciousness? And what does all this have to do with appetite? The last sentence of the previous paragraph says it all. Part of consciousness is being aware of our place in the world. We must be able to understand and then integrate instantly various sensory inputs, from understanding the meaning of words spoken to us, down to the idea of "let's eat that fruit I just saw".

Short-term buffers do not represent the entire concept of consciousness. Indeed, they do not even represent the entire concept of memory. Rather, as with consciousness, buffers are simply one ingredient in a complex definition. It must also be understood that we use these buffers in combination with other mental processes, such as the long-term memory devices we talked about in the last chapter. There is even evidence that a third memory process exists, called working memory, which serves as a liaison between the short-term buffers and long-term memory storage. We will talk more about working memory, and its role in consciousness, in the next chapter.

A summary and parting thought

We've been talking about a lot of molecules and tissues here, so let's take a few minutes for a breather and put some things into context.

As you know, we are working with the following model regarding emotional feelings: subjective emotions are experienced when two powerful collections of neurons within our bodies interact with one another. We spent most of our time talking about the first collection of neurons in our discussion of hunger, those that comprise the emotional subsystem that gives rise to our feelings of appetite. We initially looked at liver and fat tissues, and then described the role of the hypothalamus in generating appetite. We also examined several hormones whose role is to communicate between these tissues and brain regions, all in an attempt to maintain a healthy energy balance. We lastly discussed that, while a great deal of biology is beginning to be uncovered, we have much to learn before we will fully understand the nature of hunger.

As incomplete a story as it is, the subsystem is not the only aspect of appetite. The emotional experience of hunger comes when it suddenly dawns on us that the appetite subsystem has been aroused. Such budding awareness necessarily invokes the idea of consciousness, and it was to these minds-and-brains issues that we next turned. After our Descartian install-ment in the history of mind/brain thinking, we talked about another ingre-dient of consciousness, the presence of short-term buffers. We discussed how these buffers are as necessary as the longer-term ones mentioned in Chapter Two for our ability to function consciously in the real world.

We do not know much about the interactions of these two battalions of cells and their ability to cause hungry feelings in humans. Most of what we *do* know about them was not first discovered in us anyway, but rather in laboratory animals. Considering the enormous number of tissues involved, such animal work is applied only with great hesitation to humans. And *that* hints at something very important. I wish to discuss that something before we leave this subject.

Parting thoughts

Near the end of the last chapter, we discussed the fact that emotions are usually things that happen to us, rather than things we will to occur. I would like to add another characteristic to the discussion in this chapter, and, to illustrate it, I would like to return one last time to the Sixth Cornice. There is a very interesting stanza that Dante writes shortly after he encoun-ters the emaciated members of this level. Take a look at Figure 11 below, and then read the accompanying stanza.

Their eye pits looked like gem-rims minus gems
Those who read OMO in the face of man
Would easily have recognized the M.

The last two lines are especially interesting in this work, and relate direct-ly to the picture. The "OMO" comes from the quaint medieval idea that God signed His creation, just as surely as an artist might sign a painting, or a composer a score. The language God would use in his signature? Latin, of course, the words being OMO DEI, which literally means "man is of God." The ink? The very anatomy of the created thing itself, specifically the human face. The eyes form the two O's, the M is formed by the aggregate

How God signs His handiwork

A medieval idea was that The Creator, like any good artist, identified His creation with an autograph. Embedded within the face was supposedly a signature for humans, written in Latin.

OMO DEI

Dante relates that members of the Sixth Cornice were so thin that part of God's signature could be observed in their faces. The full signature is OMO DEI, meaning Man (is) of God. Dante could observe only the OMO in members of the Sixth Cornice, as shown below.

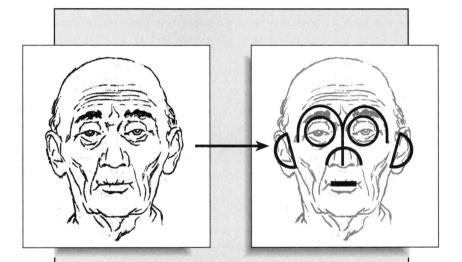

Here's how the signature is discerned:
• The eyes form the two "O's".
• A combination of cheekbone, brow and nose forms the "M".
• The nostrils form the "E".
• The mouth forms the "I".
• The ear forms the "D".

The text reads "Those who read OMO in the face of man, would easily have recognized the M."

Figure 11

structure of the nose, zygomatic arch (cheekbone) and brow, the ears make the D, the nostrils and mouth the E and I respectively. The entire concept is shown in Figure 11. Why did Dante put it here? He is relating that the starvation of the inmates was so complete that the M could be readily seen in their faces.

The idea that the spiritual origins of a biological organism would show up in physical features is a clever way to percolate theology into creation. But the idea of using anatomy to infer origins is not just a curious anecdote of medieval thinking, it is the bulwark of evolutionary history as well. Scientists still use anatomical similarities to infer origins, even coming up with ideas like "common ancestor" based on comparisons of similarities of skeletal remains. The same habit exists in the world of molecules; being able to look at the structural differences and similarities in the genes that organisms carry has greatly refined and expanded the important idea of conservation. In fact, the leptin genes were similar enough that scientists used the mouse leptin gene to actually fish out the human one. We call such a stretch the "conservation of structure". I'll have more to say about the specifics of this phenomenon in our chapter on Sloth. Suffice to say for now that conservation of structure can sometimes mean conservation of function.

The idea that conservation of structure and function exists throughout our evolutionary history plays a great role in our understanding of emotions as well. Many researchers believe that the embedded emotion-generating brain systems are equally highly conserved. They base this idea on the cruel fact that every creature on the planet has to solve a similar and continual energy crisis, finding enough fuel to live to another day. Since organisms with no appetite by definition starve to death, creating an appetite is a beneficial thing to do. Animals have evolved neural systems to accomplish this important goal, in most cases using similar looking cells tucked inside even more similar looking genes.

This similarity makes it very tempting to form a bad habit, however, and this is the point of this last section: *We look at something in animals and believe that human beings function in a similar fashion.* No question that the assumption has borne experimental fruit, but it has also produced some embarrassing dead ends, especially when it comes to behaviors. Finding a similar gene or cell in two separate species does not mean that all brains are the same, for example. It isn't that a fish writes a motet badly and a human writes one well; a fish cannot write a motet at all. Finding the way an appetite gene works in an insect is not the same thing as knowing how the gene works in you or me.

Carrying this kind of scientific responsibility doesn't erode any of our ability to understand ourselves, of course. Indeed, it is only by discovering differences that we really understand what it means to be human. Such a point would not be lost on Dante, especially as he watched how reverent the shades became when they found he was not one of them, but was instead a real live human. For them, this represents a brief flash of familiar recognition, a welcome respite from the ceaseless circling around the narrow ledges of Purgatory's Sixth Cornice. But it was only a flash, and not the real thing, and Dante moved on because he was not an inmate to a prejudice. Considering the current state of research, that's not a bad way to think of emotions either.

CHAPTER FOUR

Avarice

"When I stood on the fifth ledge, and looked around,
I saw a weeping people everywhere
Lying outstretched and face-down on the ground

'My soul cleaves to the dust,' I heard them cry
Over and over as we stood among them:
Any every word was swallowed in a sigh."

-Canto XIX, The Purgatorio

When I arrived for our morning medieval class, I noticed everybody was standing in the hall. The professor had been inside for the last hour, and only just before the bell rang did he peek his head out the door, "I want you to humor me for a second," he announced with a gleeful look in his eye. "I want you to come in, put your books against the wall and lie down, nose first on the floor." The eyes of several students widened, and the red-haired boy, dressed like he was getting ready for a medical school interview, openly declared dissatisfaction. But we all entered the room.

We found that it had been transformed. The desks had been pushed to the side, and in their places was laid a large soft rug. On the walls were giant posters of the Virgin Mary, a French Monarch labeled "Hugh Capet" and a picture of Santa Claus. Uneasy at first, one by one we lay down on the floor, nose first, eventually packed like sardines. Our uneasiness continued when suddenly one of the professor's teaching assistants burst through the door dressed like a king.

"Blessed Mary!" he shouted. "How poor you were is testified to all men by the stable in which you laid your sacred burden down!" Then another teaching assistant entered the room and declared, "I, Hugh Capet, was the root of that malignant tree, which casts its shadow on all Christendom so that the soil bears good fruit only rarely!" The professor, who had been standing at the back, chimed in too. "Time to moan everybody," he cried, "You have heard from the whip of Avarice, and are now entombed with the citizens of the dread Fifth Cornice!" Some of the students actually took him up on it, groaning and complaining about being on the floor. A few got up and looked for a desk to sit on, and then everybody rose. Eventually we were ready for class.

After the grumbling died down, the professor explained his actions. "This bit of performance art is brought to you by the inmates of avarice," with this he pointed to the teaching assistants, "the fifth level of Purgatory, where greedy people are punished. That is the subject of today's lecture." The professor then started his lecture, explaining that the Fifth Cornice was the most crowded level in the entire mountain. He also explained that the citizens were prostrate, moaning and repenting for their materialism, "Kind of like you were a few moments ago, only *their* noses were in dust, and their hands and feet were bound."

As the point of this unusual exercise began to sink in, the professor went on to describe some of the various people mentioned in the Fifth Cornice. There were positive descriptions of generosity, notably of Mary and St. Nicholas (a wealthy bishop who gave away his riches to the poor in the form of gifts, and upon whom our present day Santa Claus is based). But there were also great negative ones. Dante conversed with one of the great greedy examples, the avaricious French Monarch Hugh Capet, sire of a hoard of even greedier progeny, some of whom Dante had known.

"He began his journey with a dream, from which he derived the time of day. But as the poets moved forward, they encountered the inmates. There were both hoarders, who collected material possessions, and wasters who spent everything in life. This is the level where people with credit card debt would be interred, if such a thing had existed in the thirteenth century," the professor explained. He then related that the people of Florence could especially resonate with this level, because by the time of Dante the town was getting pretty wealthy. It had become a center of banking, and it was on the Arno, which meant it had access to the Mediterranean. The Adriatic wasn't too far off either, with the port cities of Ancona and Ravenna a short road journey away. "We meet Statius at this level," the professor said, "who as usual gives us a monologue. He describes his conversion to Christianity and his love for Virgil's poetry, a fusion of both the Roman and Christian worlds. The poets picked their way through the moaning, prostrate crowd, and it wasn't until they met another angel that they left this crowded, uncomfortable world."

A small problem in what we are going to discuss

From the exercise this creative professor put us through, the obvious subject of this chapter is the behavior of human greed. While such a biological discussion promises to be an interesting historical adventure, there's just one small problem in our attempt to be scientific about greed. The problem is that *no one has ever found a gene for avarice in human beings.* No one has ever found a region in the human brain exclusively devoted to greed for that matter. This means that either our chapter is going to be very short, or that we will have to look a little more deeply into what lies behind greed to come up with a discussion. Therefore, we may have to be psychological before we can be neurological. I have chosen the latter strategy.

The problem is how to get at the science of greed without the aid of human cells or solid definitions. When one examines a dictionary, one finds that avarice is an idea with so many different facets that a definition that makes sense to a language major, let alone to a test tube, is almost impossible. It appears that no one has stumbled upon a standard, monolithic, universally accepted definition of the concept of avarice. Because of greed's conceptual defiance (and getting a definition is the first crucial step in any scientific endeavor), we may have to content ourselves that avarice is a subjective concept in a futile search for a neuron.

It is actually enormously important to understand this lack of scientific underpinning when talking about the biology of emotions and behavior. The concept of greed, a friend to this discussion, will help us clarify the following fact: a subjective category constructed by humans does not always reflect a biological function constructed by nature. In the end, concepts such as greed may just be convenient words people use to describe aspects of the human experience, rather than real functions embedded in the brain. We visited part of this idea in our discussion of lust. As you recall, some academic professionals do not believe in the concept of emotion at all!

So what do we do?

Whether we listen to the ivory towers or not, lacking a definition that is meaningful in terms of neurons does not mean that greed ceases to exist. Our consuming need for physical possessions has historically turned sincere religious leaders into hypocrites, gentle nobleman into despots, and benign villages into dark empires. Indeed, avarice has been the central factor in many (some might say all) the turning points of human history.

Considering its importance, psychologists have valiantly braved the nebulous world of avarice and attempted to define its underpinnings. They have found some interesting footholds. Many believe that greed's underlying framework is naked human ambition (aggression), a paralyzing developmental fear (anxiety), or combinations of both. Adlerian psychologists have even tried to create a psychopathological category for people who display overwhelming greed. It is called the pleonexic personality. In their view, greed comes from fixation at the so-called anal stage (whatever that is) of emotional development, symbolically connecting feces with money. A person with such a syndrome is thought to possess (1) no sense of boundaries, (2) a distorted sense of priorities, (3) emotional detachment, and (4) general insensitivity. The pleonexic personality may also be associated with a loss of gratification.

Whether you believe in Adlerian psychology or not, a common thread appears to wind through most of the psychological thinking regarding human greed. This common thread will actually form the basis of our discussion of a particular emotion subsystem, a thread so rich in neurobiology that we can describe it in a scientifically responsible manner. We are going to talk about the biology of human *fear* in this chapter. To understand what many psychologists believe the link between greed and fear to be, consider the following true story about a man born not several centuries ago, but only a few decades back.

The tale of the greedy grandfather

In graduate school, I had an American friend whom I will call Dan, who had a Portuguese grandfather I will call Pedro. This grandfather lived an absolutely fascinating, if quite sad, life, and died at the age of ninety-four. Here briefly is his story.

Pedro grew up poor and completely disgusted with it. He was also mildly afraid of swimming in the water, which was unfortunate because he lived right on the ocean, in the seaport of Lisbon. Even without this emotional baggage, his growing years were hard; when fourteen years old, Pedro decided to leave Portugal to seek his fortune in the New World. He gathered his courage, stowed away on a boat headed for New York, and wasn't discovered until several days out of port. The deckhands threatened to throw him overboard, which completely terrified Pedro, metamorphosing his fear of water into a permanent, lifelong phobia.

While the rest of this grandfather's story is a rags-to-riches tale, it has only a form of happy ending. Pedro came to the United States, settled in Chicago, and eventually made a large fortune in the baking business. He got the money at first by being thrifty, then by playing the stock market and various security exchanges. Pedro's penchant for penny-pinching did not change as his fortune grew; indeed, his greed seemed in many ways to worsen. He refused to use his money to assist any of his other struggling still-impoverished relatives, not in Portugal, not in the United States, not anywhere. He spent very little on medical care for his wife or his two sons. Such an attitude earned him the nickname "Scrooge" to his family and, indeed, his greatest concern seemed to be with his net worth. Every day he would go down to the stock exchange, becoming visibly shaken if his stocks dropped by even small fractions of a point. As he aged, his anxiety over his holdings increased to such an extent that Pedro began suffering panic attacks, even on weekends when the exchanges were closed. And he got worse as he got richer. Towards the end of his life, members of his family would catch him getting food out of garbage dumpsters, tearing napkins in half to save money. This occurred even though by his seventieth birthday he had millions in the bank. Pedro never gave anything to charity, or to anyone else, and died as he began, physically impoverished, afraid, and, if his relatives' perspective can be trusted, indescribably greedy.

The subject is fear

Pedro's story is a common one. Most people do not like being poor for the same reason most people don't enjoy being in life-threatening situations. The common thread, winding not only through this grandfather's story, but also through the idea of a pleonexic personality, is human fear, one of the two great motivations psychologists say undergird greedy people. Pedro certainly experienced a lot of fear: phobia, general anxiety, perhaps even post-traumatic stress. While there isn't much to say about the biology of avarice, there is great deal to say about the biology of fear and anxiety. Since we are going to talk about aggression in a later chapter, we shall focus on human fear and anxiety in this one.

We will start our biological description by talking about five regions of the brain known to be deeply involved in human fear and anxiety, and then look more intimately at some of the genes involved in fear's neural subsystems. With this information under our belts, we will extend our twin discussions of the history of mind/brain issues, and add another ingredient to our definition of consciousness. Specifically we will focus on one aspect of consciousness without which fear could never be experienced: the idea of the so-called working memory. Finally, we will return to our ideas about human categories versus biological functions, and relate them specifically to our link between the notion of avarice and the fear that appears to undergird it.

Getting into the biology via a painting

The fear we feel comes from a fairly well-defined underlying set of processes. But fear can have so many different definitions that a legitimate question can be asked: which fear are we discussing? Take the various anxieties of my friend Dan's grandfather. One might intuitively understand how Pedro's fear of poverty – a survival reaction to a perceived threat – might make him greedy. But he also had a fear of swimming, which seems more like a phobia than a fight-or-flight response. Pedro also experienced anxiety attacks, not brought about by any external threat, but only by internal fear. How far does fear go before it becomes so pathological that people like Pedro become greedy misers? And how do we organize it all once we know?

We get clues about how to categorize the concept of human fear from certain psychiatric definitions. Mental health professionals call pathological fear anxiety disorders. Subcategories include phobias, panic attacks, PTSD (post-traumatic stress disorder), OCD (obsessive-compulsive disorder) and generalized anxiety. We have learned a great deal about normal fear by studying these pathologies, and we will talk about several of them shortly. However, before we get started, we need to address some important questions, considering our previous discussion of human categorization versus biological reality. Are categories of fear governed by separable biological systems? Are they part of the same thing? Since there are no genes of avarice, might there be no genes of anxiety either? These complex questions have, as you might expect, complex answers. To understand how many biologists currently view the underlying biology of fear, we will need to use an analogy. Consider the following illustration, taken from the medieval world of altar-piece painting, describing a work that actually survived Donati's sack of Florence in the early 1300s.

When Dante was a year old, a child was born in his hometown who would grow up to be as famous as the author. The little boy was the great Italian painter Giotto, whom Dante actually mentions in one Cantos of *The Purgatorio*. Why did Giotto become famous? Among other reasons, his celebrity had something to do with his style. The characters in a Giotto painting are a revelation in gravity and concentration. Giotto made sure that his people possessed expressive faces, unlike many of the paintings of his contemporaries. He did this through his use of light, creating stunning three-dimensionality to the anatomies of his characters.

One of Giotto's more interesting works lies in the National Gallery in London, a painting executed on wood and called *Pentecost*. Rich with those expressive characters, the panel is supposedly one-fifth of a larger work, all painted originally on a single plank of wood and used as a decoration behind a medieval Italian church altar. Depicting various scenes of Christ's life, four other panels have been scattered across the world and are now located in museums in the United States, Germany and Italy.

Though the styles are all similar – and one might intuitively make the connection with Giotto based on execution – art historians used to argue whether these five now-scattered panels were painted on a single plank of wood. To answer that question, one would have to do the unthinkable: scratch off pieces of the painting, examine the underlying grain of wood in each picture and start looking for linear connections. Fortunately, we do not have to perform such artistic butchery to get our answer. Modern X-rays have been aimed at each panel, and, because of an X-ray's peculiar nature of passing through some objects but not others, the underlying grain of wood has been revealed. Sure enough, Giotto painted his five panels on a single plank of wood, beginning with the panel now residing in the Metropolitan Museum in New York, ending with the panel now hanging in the National Gallery in London. I have had the distinct privilege of seeing the *Pentecost* panel in London, and it is indeed a powerful study in expression and light.

So, what does such a vivid, now X-rayed, painting of a fourteenth century Italian master have to do with the biology behind the human experience of dread? I use this illustration because of the various human categories of fear we mentioned previously. Anxiety, phobias and fight-or-flight feelings may seem to be different things, only as distantly related to each other as New York is to London, a Nativity to a Pentecost. But when one looks underneath the human categories of fear to the underlying biological roots, one sees a strikingly common biological grain. A specific set of tissues, cells and genes appears to form the basis of most of our fears, regardless of the human category we ascribe to subjective experience.

Unlike the plank of wood, however, that fact of commonality does not mean the biological subject is linear or straightforward (few things in biology are). The cells and tissues that form the sensations of fear are as varied and far-flung as the museums exhibiting Giotto's altar-piece. Indeed, a great deal of research has been done to elucidate the biological basis for the emotion system of fear, and it is to these data that I now turn. I will attempt to show that descriptions such as avarice are manufactured categories, revealing not specific subsystems of greed, but a primitive method of detecting danger that has now actually run amok.

Five regions of the brain

So what happens when Pedro sees a dip in the stock market, and becomes as fearful as if he just saw a snake? You already know the answer. His fear subsystem becomes activated, a large part of which resides in his brain, and this stimulation is brought into his awareness. Five major regions under the skull are involved in the processing of fear, and I would like to briefly describe the function of each before we talk about how fear occurs (see Figure 12).

The organ of fear

Among the many tissues in the human brain, the region called the amygdala is the one most responsible for feelings of fear. The amygdala interacts with a large number of other regions in the brain, as shown below.

Many regions associated with threat responses in the brain provide inputs into the amygdala. In turn, the amygdala is connected to still other regions in the brain responsible for generating fear responses. As can be seen in the drawing, a region in the amygdala known as the central nucleus helps provide many of those output connections. Other regions of the brain analyze characteristics of the threat response.

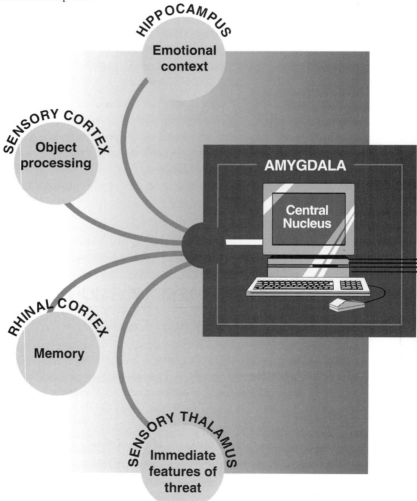

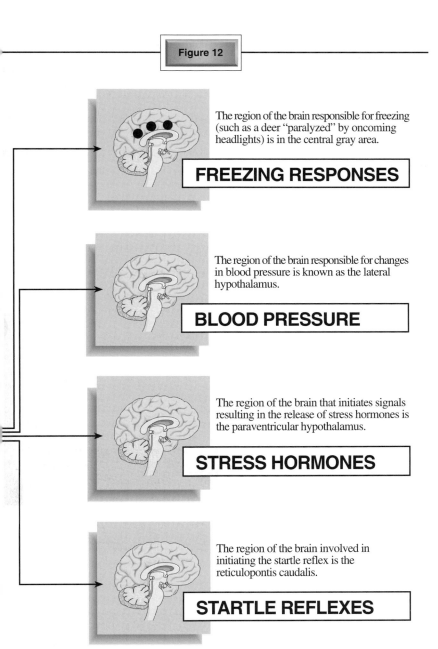

Figure 12

The region of the brain responsible for freezing (such as a deer "paralyzed" by oncoming headlights) is in the central gray area.

FREEZING RESPONSES

The region of the brain responsible for changes in blood pressure is known as the lateral hypothalamus.

BLOOD PRESSURE

The region of the brain that initiates signals resulting in the release of stress hormones is the paraventricular hypothalamus.

STRESS HORMONES

The region of the brain involved in initiating the startle reflex is the reticulopontis caudalis.

STARTLE REFLEXES

Thalamus. This region of the brain acts like an air traffic controller at a busy airport. Many different kinds of signals are routed through it. There isn't one signal from the sensory tissues (for example, our ears or eyes) that doesn't connect in some form to the thalamus. In turn, the thalamus has connections to the higher centers of our brain, such as the cortex. The hypothalamus, already discussed, literally means "under the thalamus", and indeed the thalamus sits atop of it like a roof.

Amygdala. Perhaps the most important fear-generating region of all, the word amygdala literally means almond. This tiny organ, also located in the middle of our brains, has a great deal to do with generating and mediating certain kinds of emotional behavior. One particular region embedded in the middle of the amygdala, called the central nucleus, appears especially to be concerned with the generation of fear. Destroying this region in laboratory animals inhibits nearly every outward physical manifestation of the fear response, from stress-related hormone release to the upregulation of the heartbeat and changes in blood pressure (we'll talk more about those later). The amygdala, once stimulated, can send messages to other areas of the brain, including the familiar hypothalamus.

Hippocampus. Literally "horse-shaped sea-monster", the hippocampus also sits in the middle of the brain and controls a wide variety of processes. One of the most important has to do with remembering things. If a person's hippocampus is damaged, they lose certain memory abilities. This includes being able to remember what they did recently, as well as the ability to both store that memory and later recall it when needed.

Cerebral cortex. Literally meaning "bark", the cerebral cortex is a layer of cells blanketing the hemispheres of the brain. It is often called the gray matter. In our discussion of fear, we will be talking specifically about cells in the sensory cortex. These are regions of the cerebral cortex involved in the reception and interpretation of various sensory inputs. Examples include the visual cortex, involved in the integration of visual input, and the auditory cortex, involved in integrating sound.

Pituitary gland. This little appendage hangs off the hypothalamus like a Christmas ornament. It performs several executive functions, chief of which is the up- and downregulation of certain hormones. When a creature encounters a fearful situation, the pituitary gland is instructed by other regions of the brain (the hypothalamus, for example) to release certain stress-related hormones.

Like Giotto's altar piece painting, these five regions can be thought of as a single biological plank upon which our feelings of fear are painted. There are, of course, many other areas of brain tissue involved in generating the fear response. Indeed, understanding how these five regions interact both with each other and with the rest of the body will go a long way towards explaining the emotion of fear. It is to the fascinating interaction of these various regions that we next turn.

Interactions and the feelings of fear

So what happens when we detect danger? You might well imagine Dan's greedy grandfather sitting like a quivering deer in front of the Chicago Mercantile Exchange, anxiously searching the boards for his riches. What happens to Pedro's body when he suddenly discovers he is losing money? To hear Dan describe it, Pedro might as well have been looking face-to-face at a growling bear, poised to attack him. Exactly what occurs in Pedro's mind the instant he sees a minus sign on the reader board?

The answer is that information about potential danger is trafficked along both a fast and a slow neural road deep in Pedro's brain. His eyes first see the negative sign and immediately the signals flash through his eyes and deep into Pedro's head, arriving at the air-traffic-controlling thalamus. As we mentioned, this organ processes inputs from many different signals, including those coming from the eyes. In response to the negative sign, the thalamus activates both the fast and the slow neural signal pathways.

Just what exactly happens on the fast road? The fast road is a group of neurons that bypasses the rest of the brain and goes directly to the fear-generating amygdala. The signal is coarse and unrefined because it has not yet been processed by the more scholarly parts of the brain, namely the sensory cortex we described previously. But speed is always of the essence in a potentially dangerous situation, and so the fear generator is alerted first, even before we fully know the magnitude or nature of our troubles. As a result of stimulating the amygdala, Pedro's body begins to physically react in a classic fear response – his heart beats more quickly, his blood pressure changes, and so on. Indeed, it is possible that this route is responsible for creating emotional responses we do not currently understand.

Bypassing those higher centers means that we may only be dimly aware of the threat at first. Does it sound odd that we begin responding to danger before we really become aware of what is dangerous? We have actually run across this consciousness-bypass routing before. You recall from the lust chapter that many of our emotions appear to lie outside our conscious awareness, and that the ones we feel are those that collide with our awareness abilities. This is just another example that emotions are something that happen *to* us, rather than activities we will to occur.

I mentioned that there is also a slower route activated by the thalamus as soon as a potential danger cue is sensed. What is that all about? In addition to directly stimulating the fear-generating amygdala via the short road, the thalamus generously routes some of the danger signal to the higher centers of the brain, namely the sensory cortex. As Pedro's eyes receive input from the stock market boards, the information is tossed to the part of the cortex involved in visual processing. Once the signals arrive in this region, the brain gets a moment to think about what it has just encountered. It creates a truer, more thoroughly detailed representation of what was observed. Such ruminations take time, which explains why it is the slower road. After giving the input a more detailed treatment, the cortex makes up its mind and sends signals back to the amygdala, the same destination as the fast road. Only now the signal is more precise and Pedro starts to become aware of his feelings of fear. We will talk more about the routes taken by the sensory cortex shortly.

What's the big deal about the amygdala?

You probably noticed that whether the signals took the slow or the high road, all the information about fear eventually came back to the amygdala. In fact, they are routed to a specific area in the amygdala known as the lateral nucleus, which you can liken to the amygdala's front porch. From there, the signal is transmitted to the rest of the organ and the various bodily reactions to fear can occur (like the so-called startle reflex, freezing, elevated heart beat and blood pressure, etc.). These data underscore the fact that the amygdala plays the central role in our response to dangerous situations. But what is so special about the amygdala that it mediates such an important response? And how can signals arriving at the front porch of one organ do so many disparate things? The answer comes in part from comprehending how the signals are processed once they arrive at the amygdala. It turns out that the amygdala is a house possessing many separate "rooms" besides the front porch of the lateral nucleus. One such area, itself a complex mass of neurons, is called the central nucleus. When a fear signal arrives at the central nucleus, a whole bunch of things begins to happen. This complexity requires an illustration to understand its role in fear most fully. Let us return to the world of the Fifth Cornice and focus upon an interesting medieval view, not of the inner world of the body but of the outer world of the sky.

Astronomy and a brain

The Canto XIX in *The Purgatorio*, the one in which the poets first encounter the Fifth Cornice, begins with a very unusual passage. Dante is about to describe the time of the day he began dreaming an important dream, and he uses planetary references to establish the time.

> At the hour when the heat of the day is overcome
> by Earth, or at times by Saturn, and can no longer
> Temper the cold of the Moon; when on the dome
> of the eastern sky the geomancers sight
> Fortuna Major rising on a course
> On which, and soon, it will be drowned in light:
> There came to me in a dream . . .

Though it may not seem obvious, Dante is actually describing the special hours just before morning, a time when people supposedly dream only of true things. Why in the world would Dante use such odd-sounding language? The answer comes from understanding how medieval people viewed certain astronomical events, an understanding that provides an analogy for the processing power of the amygdala.

In Dante's time, it was felt that sunlight could provide two separate climates, depending on where the reflected shafts fell. If sunlight fell on the Earth, for example, it was reflected to the Moon. This reflection was felt to produce warmth on the lunar surface. Conversely, if the sunlight fell on the Moon, it was reflected back to the Earth. This reflection was felt to produce cold on the terrestrial surface, hence the reason why the air was chilled at night. Since the accumulated heat of the day would be gone by the time Dante began his dream, it could no longer "temper the cold of the Moon"; hence the poet is talking about early morning. The reference to Saturn comes from a similar idea about reflection. Saturn was thought to be a cold planet, reflecting back cold light (Mars, on the other hand, was thought to be a hot planet, reflecting back warmth). This medieval astronomy shows the many diverse reactions one can get from a signal emanating from a single source.

What does all that have to do with the central nucleus of the amygdala and our responses to fear? Just as the medieval Sun can create disparate reactions by tossing its photons to various reflective surfaces on the planets, the central nucleus can toss the electrical signals of fear to various reactive regions within the brain, and get different responses. Some of the brain's reactions to the information are varied indeed. Here's a quick summary of what occurs once the fear information has been routed to the amygdala's interior (for a more detailed explanation see Figure 12).

1. The central nucleus can route fear signals to a region responsible for the "freezing" response.
2. The central nucleus can route fear signals to a region responsible for changes in blood pressure.
3. The central nucleus can route fear signals to a region responsible for the release of stress hormones.
4. The central nucleus can shuttle signals off to a brain region that, when stimulated, initiates the startle reflex.

This is something like the differential effects of a single medieval Sun shining on distant planets, with a variety of messages reflecting away from the central nucleus even though it only received a single signal. Taken together, this processing is how the amygdala helps mediate the fear response. Even though the central nucleus is receiving a common set of danger signals, it is able to reflect the information to various regions of the brain to produce the widely disparate reactions we call fear.

The remembrance of things past

Examining the biology of the amygdala is not the whole story behind fear. We have not yet addressed how the emotion is generated in the first place, nor have we discussed the role of memory in the creation of fear responses.

What do I mean by memory in fear responses? To understand how yet another organ in the brain works in the creation of the fear response, let us revisit Dan's grandfather Pedro for a moment. Pedro's greed appeared to be fueled by a number of fears, several of which appeared to involve his ability to remember things. As a young man, Pedro developed a terror of swimming, a fear that turned into a permanent phobia after his experience on the cargo ship. His meager beginnings bequeathed to Pedro a life-long fear of poverty, probably fueling much of the greed his relatives encountered. Pedro even developed panic attacks, which did not appear to be caused by any specific stimulus, occurring even on weekends when the stock markets were closed.

Though water, poverty and weekend panic attacks may seem very different kinds of fear experiences, they all have two things in common. Every one of these experiences involves the use of specific kinds of human memory, and none of these fears requires an *immediate* threat to be experienced. It is this distinction from the acute fight-or-flight feelings induced by a snake or a stock-market meltdown that I wish to explore next. As they are involved in memory, they will all make use of the hippocampus, one area of the brain involved in human memory.

The issue is anxiety

When we view Pedro's fears (the ones that aren't a reaction to an immediate threat), we are really talking about the psychological category of anxiety. While it is true that anxiety and fear are closely related, they are not exactly the same thing. The immediacy characteristic described above is the chief difference. Pedro could become anxious about swimming just by remembering his unpleasant experience on the boat, even though an ocean might be miles away. Some people believe that anxiety is nothing more than fear without a closing chapter – in psychologist's parlance, unresolved. If an immediate, terrifying situation cannot be escaped or avoided, the fear transforms into anxiety.

Fear and its remembrances are not abnormal aspects of the human condition. It is only when these feelings are experienced more persistently than circumstances call for that the mental health professionals break-out their notepads and start diagnosing. Psychiatrists call the abnormal kinds of input-independent experiences anxiety disorders. They are divided into five categories: phobias, PTSD, OCD, panic, and generalized anxiety. As you can see, Pedro appeared to be in the grip of several of kinds of anxiety-related disorder.

While these categories represent attempts by skilled clinicians to organize and then treat human suffering, it is quite possible that all anxiety disorders exploit the same brain systems used in typical fear responses. The only distinguishing feature of anxiety disorders appears to be the use of memory. Exactly what role does memory play in the experience of such disorders, and which areas of the brain are responsible for its creation? Answering these questions will go a long way towards our comprehending the emotion of fear. To understand the answers, we need to work a little more closely with definitions of human memory, and then return to the familiar amygdala, and its relationship to the hippocampus.

Memories and brain cells

In this and the two previous chapters, we have been discussing various categories of memories in our quest to more clearly understand the ingredients of consciousness. The brain has many more memory systems than just long-term storage, short-term buffers, and working memory. There are two other categories of memory I would like to discuss here, each apparently mediated by different regions in the brain. Understanding these categories will help us explain the link between our memories and our fears.

The first category of memory is often called declarative or explicit emotional memory. This is mediated by that sea-horse-shaped hippocampus, and its various connections to higher cortical areas. The other category is called implicit emotional memory, often called unconscious memory. This form is mediated by regions other than the hippocampus, among them our familiar amygdala. Both explicit and implicit memories cooperate to create anxiety as a direct response to fearful situations. Here's how they are thought to work:

If Pedro is first exposed to a traumatic situation, say the threat by the deckhands to throw him overboard, both the explicit and implicit memory systems in his brain are activated simultaneously. The implicit memory system is involved in the immediate physiological response, the explicit memory stores the details. If Pedro goes below deck after the threat goes away, but then later returns topside and sees the ocean, these memory systems carrying the traumatic event can be re-activated. As a result, Pedro may experience severe anxiety. The hippocampus will remember who gave the initial threat, where on the boat Pedro received the threat, the time of day it occurred and, perhaps most importantly, the fact that the event was terrifying. The amygdala, working through its powerful central nucleus, will cause Pedro's heart to race, his blood pressure to change, and stress hormones to be dumped into his bloodstream, in fact, all the responses associated with the initial fear. The interesting thing about anxiety is its total dependence on memory. As Pedro stands on the deck later that evening, there is no one threatening to throw him overboard. He is experiencing anxiety as a direct result of his ability to *remember*. This remembering occurs because the hippocampus gives explicit details about the situation, and the amygdala, using implicit emotional systems, causes Pedro's body to quake.

To be sure, it is easy to think that these two memory systems, because they work so closely in parallel, are part of a single memory–fear system. Though their cooperation is extreme, they are actually distinct entities. It took a great deal of animal experimentation, and a painstaking search for human subjects who had experienced a particular kind of brain trauma, to discover the presence and biological tissues of implicit and explicit memories.

This overboard experience is something that Pedro never forgot. Indeed, any of us going through such an experience might carry the same permanent recollection of the experience. For Pedro, it reinforced a fear that eventually blossomed into a full-fledged phobia. Do scientists have any clues as to why these memories can be so indelibly etched onto the neurons of our brain? Can the mechanism that allows phobias to exist in humans be explained from an evolutionary perspective?

The answer to both questions is "possibly"; there is a great deal of speculation as to why the human brain has a capacity for anxiety disorders, some of which make a great deal of sense. In a relatively stable environment, for example, one is not likely to encounter brand new threats to life every minute (if you are living in the African Savannah, for example, you are not likely to run into a polar bear). An encounter with an *unknown* deadly force, however, might quickly be lethal because of its rarity. A new poisonous snake could enter the bush country and wreak havoc on a human population simply because nobody knew it was deadly. Thus, it is useful to have the mental capacity to remember instantly when one comes across a new threat, and then to generalize it, getting ready for even novel encounters. Your brain might store the notion, "Watch out! New snakes may be bad, so be careful every time you encounter a new snake!"

That's a useful mental talent to have as long as one is in the bush country. But we live in a very different environment in the twenty-first century, and it is this change in context that might turn a perfectly logical survival strategy idea into an irrational pathology. For example, cars are indispensable machines in many parts of the world. Regardless of their utility, the tendency may be to fear getting into one the day after an accident. If the accident is bad enough, the fear may turn to anxiety. The problem is that cars, unlike poisonous snakes, aren't inherently deadly, and we say that the fear of getting into a perfectly safe car is irrational, even phobic. One way to explain the anxieties people experience is to look at the historical context of our need to handle threat experiences. It all has to do with memory, which is why we mention it in a chapter on the emotion of fear.

Sensory cortex

You recall in our discussion of the amygdala that fear-response information takes both fast and slow roads through our brains in order to alert us to danger. The reason a slow road exists at all is because the information is shuttled through the sensory cortex before returning to the amygdala. The fast road goes directly to the amygdala, no fancier processing involved. Having discussed the hippocampus, it is now clear that memory is deeply involved in the processing and experience of fear. As you might suspect, the region is deeply involved in the slow road, and possesses connections to and from the sensory cortex. The routes and roles surrounding this sensory cortex in relation to fear are what I would like to briefly discuss next.

There are regions in the brain we have not yet discussed; for example, the temporal lobes, a massive group of neurons that drop down like large earrings on both sides of our brains. The temporal lobes, of all things, possess a kind of memory system. What does that have to do with fear? If the temporal lobe is removed from monkeys, an absolutely astonishing thing is observed. The animals lose their fear of things of which they should be – and normally are – terrified. They will approach unfamiliar humans and even other unfamiliar animals unfalteringly and without hesitation. They tend to examine things by mouth, rather than by hand, seemingly unable to recognize objects by sight. Even with the oral stimulation, they show no signs of fear, freely putting screws, sticks, pieces of apples, live rats, and even live snakes into their mouths. The objects are discarded only if they prove not to be edible.

What in the world is going on with this temporal lobe? And how does it all relate to fear, or rather to its loss? The experiment just described inaugurated a powerful research effort that uncovered the role of memory and its association with those sensory cortexes we were just discussing. As you might expect, the memory-mediating hippocampus plays a great role.

As you know, we can see and smell and feel things because of complex interactions with our senses and the regions in the brain to which they are connected. These regions are sometimes referred to as the neocortex.

There is a visual cortex, which helps us visually, an auditory cortex which helps us aurally, and even a somatic cortex, which helps us in our response to touch.

These regions are not isolated little dollops of paint on the neural canvas of our brains, of course. As with all regions, they are deeply connected to other regions of the brain. They are even indirectly connected to the hippocampus. Why do I say indirectly? These sensory islands send out projecting neurons to an intermediate region of nerve cells, and it is these intermediates that are connected to the hippocampus. The hippocampus and the neocortex can talk to each other, plainly enough, but they have to go through the transitional cortex to perform their communication. The neocortex and its associations with the transitional cortex occur in areas of the temporal lobe. In fact, those cortexes form the temporal lobe memory system.

So what does all this have to do with fear? A great deal of experimental work has been done to uncover the following pathway:

1. The sensory cortex receives information about the environment from the sensory organs. The sensory cortex processes this information, creating perceptual representations of the inputs. This is the slow road.
2. These representations are shuttled to the transitional cortex, which then sends them to the hippocampus. The hippocampus processes the information, and sends its own signals back to the transitional cortex. The transitional neurons relay back to the neocortex. Memory forms.
3. The memory becomes permanently stored in the temporal lobe.

The role of fear in this interesting pathway can be easily discerned if one remembers the monkey experiments. The monkeys lost their fear of certain things probably because they lost the memory trace informing the animal of the danger. It appears that visual processing of many types was affected, since visual input alone was not enough to create fear. How important indeed are these higher cortical centers of the brain in our perceptions of danger!

The pituitary gland

In discussing the amygdala, hippocampus and cortex, it almost seems that the body's fear response is limited to what occurs inside our skulls. Nothing could be further from the truth. The brain and the body work closely together as the feelings of fear unfold. Indeed, our brain's ability to sense what is occurring in the body may contribute deeply to the emotion of terror. How does the brain work with the body to elicit the physiological response of fear?

Previously, we discussed the fact that stress hormones are involved in the fear response. Understanding how these hormones work is the gateway towards understanding the body's role in fear. The central nucleus of the amygdala works with another area of the brain to ensure that those hormones are dumped into our bloodstream. The organ that does this is called the pituitary gland and I would like to talk about it, and how its hormones work, next.

The pituitary gland secretes chemicals that are so powerful that their presence is almost like inciting a war in the body, and the principle behind the mechanism is complex enough to require an illustration. This violent feedback reminds me of a historic incident I first learned about in a course entitled "A History of the Catholic Church" back when I was an undergraduate. Dante actually alludes to this incident in a stanza of the Fifth Cornice.

> Oh Avarice, what more harm can you do?
> You have taken such a hold on my descendants
> They sell off their own flesh and blood for you.
> But, dwarfing all crimes, past or yet to be,
> I see Alagna entered, and in His Vicar,
> Christ Himself dragged into captivity.

What exactly is this incident "dwarfing all crimes, past or yet to be"? My professor described the crime as a direct result of certain inflammatory messages ping-ponging between a medieval King and an aging Pope.

"Philip IV was a jerk, pure and simple. About as ambitious a French King as you could ask for in the late Middle Ages." The class laughed good-naturedly at the teacher's editorializing. The professor continued, "And about the only guy who was more ambitious than this French King was the gent who happened to be head of the church, Pope Boniface VIII. They did not (how shall we say it?) get along like pastor and congregant."

The professor went on to describe some of the interesting communications between the monarch and the pontiff. "The core issue was greed," the professor said (probably the reason why Dante included it in the Fifth Cornice) "which manifested itself in the question of whether or not the state could tax the clergy. The King and the Pope would regularly hurl incendiary messages at each other, inflaming their passions, engendering both fear and loathing." The professor described that Philip convened an (illegal) general council, indicting the Pope as a murderer and a tyrant, complaining that he was filled with avarice and ambition, unfit to be the Vicar of Christ. The Pope was informed that such a council was convened, and promptly sent a message from his palace at Anagni, telling Philip that only the Pope could convene councils. "And then Boniface did the natural control thing," the professor continued. "He prepared papers to excommunicate Philip, and made plain to the King that such papers were being drawn." The professor then described one of the blackest days of the papacy. "That excommunication message was apparently all that Philip would stomach. He dispatched two thousand mercenaries to Anagni, sacked the palace and took the Pope prisoner for three days! The action effectively ended the excommunication order. It was a short-lived invasion, fortunately. The townspeople ran the mercenaries out of town, reclaiming the palace, but not before exacting a terrible toll on Boniface. He soon fell ill with a fever and died a few weeks later." Though Boniface himself was no angel, this capturing of the Vicar is what Dante calls the dwarfing crime of all the ages.

But what about the hormones?

I bring up this medieval series of fear-engendering messages flowing between two power centers in our chapter on stress hormones for a specific reason. In many ways, the incident between Philip and Boniface parallels what occurs in our bodies as the brain experiences fear. The brain is one power center, and a specific gland located in our mid-abdomens, called the adrenal gland, is the other. Here's what I mean.

When the brain experiences a stressful situation, the amygdala sends messages to the hypothalamus, that interesting organ we discussed in our chapters on lust and gluttony. The hypothalamus doesn't keep the information to itself. Rather, it tosses the fear signals to the pituitary gland. Much of this communication is mediated by specific hormones, chief of which is known as CRF (corticotrophin-releasing factor). The pituitary gland is designed to respond quickly to CRF. This stalwart secretory tissue is responsible for many other hormones that communicate with the rest of the body.

The CRF consultation results in a series of action steps for the pituitary.

The pituitary orders the release of another hormone known as ACTH (mercifully short for adrenocorticotrophic hormone). This hormone travels via the bloodstream to the other power center, a group of tissues called the adrenal gland, which sits on our kidneys literally like the roof on the cathedral at Chartres.

ACTH is a powerful signal when ushered into the presence of the adrenal gland. As a result of ACTH's message, the adrenal gland marshals the creation of a host of hormones, an army of chemicals scientists call adrenal steroid hormones. These hormones march out of the adrenal gland, enter the bloodstream and do lots of things, such as helping the body to mobilize its energy supplies to deal with the threat. Eventually, these hormones storm the neural halls of the brain, targeting both the amygdala and the hippocampus. What happens when these hormones break into the brain, especially the hippocampus? They shut down the release of CRF, the hormone that told the pituitary gland about the stressful situation in the first place. One of the consequences of this inhibition is to help dampen the fear response in the creature undergoing the stress. In many ways, the interaction is reminiscent of Philip's assault on the Pope's palace, resulting in the destruction of the excommunication order.

The biochemical story doesn't end in brain signal inhibition, of course. It doesn't even end with the feelings of stress being destroyed. If the threat continues, more CRF is released to the pituitary, more ACTH reaches the adrenal glands, and more steroid soldiers storm the brain. As long as the stress continues, the body will continually balance two sets of signals, the excitatory inputs from the amygdala, and the inhibitory inputs coming to the hippocampus.

Curiously, if the stress continues for too long, the memory functions of the hippocampus begin to flicker off. This faltering can be seen behaviorally. As the hippocampus is so involved in memory and learning, stressed animals become increasingly disoriented and confused, unable to learn and remember specific laboratory tasks. In humans, stress most affects our explicit memory functions, a deficit that can even be seen anatomically. People suffering from PTSD have a shrunken hippocampus. There is no loss of IQ, or other kinds of mental functioning, just significant memory impairment. There is evidence that this shrinkage occurs because of the continued presence of some of the steroid hormones mentioned previously. Curiously, the function of the amygdala does not seem to be impaired by exposure to long-term stress.

A summary of five tissues

These five widely dispersed areas of the brain, much like Giotto's altar piece, work together to accomplish a specific goal. These neurons and attendant body processes are the substrates upon which we experience the various hues and shades of fear. With so much complexity and so much at stake, one might first expect to see some kind of extremely efficient routing system in the exploration of these substrates. Indeed, the thalamus provides for us exactly that, allowing both fast and slow roads to exist in our evaluation of various threats. One might also expect that the router would eventually connect to some kind of fear-generating device, one capable of marshalling many resources in the attempt to survive hostile situations. That is exactly what we find in the amygdala, whose powerful central nucleus commands all kinds of fear responses in the body. Since an accurate memory of a given threat is useful to have if a creature is going to survive beyond an initial stressful encounter, one might expect to see sophisticated connections to storage devices. That is exactly what one sees in both the neurons in the hippocampus and in higher level devices in the cortex. Since other regions in the cortex give us information about what our senses are perceiving, the entire fear apparatus has to be able to both connect and respond to our "thinking" regions. Otherwise, like those poor monkeys, we will never be able to separate out what is truly dangerous from what is truly inert.

We still have a long way to go before we fully understand all the many shapes and colors of human fear. Even the explanations above are really only a cursory treatment of what is known. And we really don't know very much, intimidating as that may sound. What we really need is the functional equivalent of an X-ray machine, one that can penetrate the layers of tissues and see the common mechanisms available throughout the brain, ones separating anxiety from phobia, PTSD from OCD, and so on.

It is very possible that we have recently come across just such a series of techniques. The powerful tools of molecular biology are beginning to uncover some of the mystery inside the neurons that govern our fears. I am talking, of course, of the world of the gene, a world we have visited in every chapter so far. I would like now to talk about the isolation of certain proteins and gene sequences that play a powerful role in the generation of fear. The discussion will be frustrating, of course. Though these techniques are revealing data that truly lie beneath the surface of the brain areas we just discussed, we will find that the information raises just as many questions as answers.

The genes of fear

Figuring out the wiring diagrams of the human brain is a task so daunting that some researchers used phrases like "it may be impossible" to describe the effort. Some have gone so far as to say that looking at the connections is a waste of time, since stimulating one area of the brain can influence so many other kinds of neurons.

As a graduate student, I once attended a lecture from a guest neurobiologist, a guy who had spent his life looking at these wiring diagrams and the genes that guide their placement. He was talking to a general public audience, and lectured with almost manic enthusiasm. "There are billions of neurons in our brains. Billions. More neurons than there are stars in Carl Sagan's galaxies." As hardened graduate students, we tried not to be spellbound, but it didn't work. He continued, "And if that's not enough, a single neuron can burst into a thousand different branches – and all those branches connect to other neurons. Pull out all the neurons in a pea-sized area of the cortex, lay 'em end to end, and you will wind up with a line almost two miles long! And that's just in one area of the brain!" He said one last amazing thing, "In terms of connections, do you wanna know who the record holder is?" And of course we did. "The record holder", he declared, "is a Purkinje cell, a kind of nerve cell buried deep in the brain. It can have connections to maybe one hundred thousand other neurons! Now you go home and try to figure out how we are going to understand something as complex as a Purkinje cell!"

With all these bewildering set of connections, it is a surprise that anybody has taken the task head-on, trying to understand how the pathways lead to behaviors. And the funny thing is, there is just as great a complexity to behold if you simply open up the hood of a *single* neuron and peer into its internal genetic make-up. Of course, looking at the molecules that drive these dense collections of neurons would give us the most intimate description of their function. And so, a few brave souls like this lecturer have plumbed the depths of their intellectual courage, spent eight to ten years of their lives learning the techniques, and have become researchers into the world of the neural gene.

And just exactly what have they found? Especially as it relates to the feelings of fear? We discovered, long ago, that neural cells communicate with each other electrically. One neuron is electrically stimulated, and has the ability to pass the stimulation to its neighbor. The question of interaction then becomes a question of "how is this communication achieved?" To understand how molecules work in neural communication, creating these interesting networks, it is important to review a few things about neurons, their overall structure, and their interactions. Then we can talk in an understandable way about the genes that govern their behaviors.

An analogy in commerce

A neural cell looks kind of odd when viewed under the microscope. It reminds me of what a mop might look like if it had just been scared out of its wits. At one end lies a frenzied tangle of fronds, which scientists have named dendrites. These dendrites are attached to a sphere-like structure that researchers call the cell body. This roundish structure contains the cell's nucleus, which, as you recall, is the bag that holds the neuron's genetic information. The cell body is itself attached to something that looks like a pole, a long stick-like structure scientists term an axon.

As the guest lecturer stated, there are billions of these things crammed into our skulls, all chattering with each other in electrical ecstasy. Exactly how one neuron stimulates another to pass along its information is complicated enough that I would like to return to the world of the Fifth Cornice and discuss another medieval analogy. This one involves the city in which Dante and Giotto were born, the Italian city of Florence.

Florence was an interesting place to be in the thirteenth century. Lying on the Arno river, it had access to the Western Mediterranean even though it was an inland town. There was also indirect access to the Adriatic, via overland trips to the ports of Ravenna and Ancona. This meant trade was possible with Balkan ports like Zadar (in present-day Croatia), as well as points east, far into Asia. Thus Florence had access to both Western Europe and to the east in Dante's time. The city, as you might expect, became wealthy, famed for its fine textiles, silks and wool, dyed in a secret process zealously guarded by skilled craftsmen. Florence also became a famous banking center, attracting merchants and businessmen from around the Western World.

The analogy concerning how neurons communicate involves a hypothetical situation involving a textile supplier in Florence and a retailer in Croatian Zadar. Suppose that the textile supplier in Florence wanted to send a sample of his wares to a merchant in Zadar, in hopes of generating a trading relationship. As roads connecting latter-day Yugoslavia to Italy were either poor or nonexistent in the fourteenth century, the Florentine supplier would probably look to ocean-going vessels for his communication. Since the seaport at Ancona lay directly across the Adriatic Sea from

How neurons communicate

Using tiny molecules known as neurotransmitters, information can be transferred from one neuron to another. Here's how it works.

"Connecting" neurons are not really connected at all, but are separated by a space called a synapse. Such interacting neurons are faced with the daunting task of crossing the synapse to communicate information. This is done with mobile neurotransmitters, which are stored in bags called vesicles inside the nerve cell until they are needed (see below). Facing the cell with the neurotransmitters, the neurons on the other side of the synapse possess a class of molecules known as receptors. These receptors, tethered to the cell's surface edge, project outward into the synapse. Receptors have two jobs: (1) they bind to neurotransmitters and (2) they tell the cell to which they are anchored that a neurotransmitter has been received.

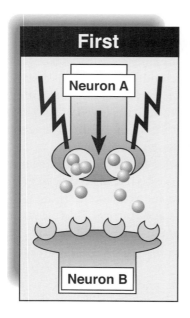

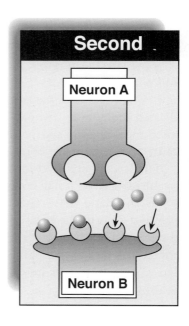

STEP 1: RELEASE OF NEUROTRANSMITTER	STEP 2: BINDING TO A RECEPTOR
Neuron A becomes electrically stimulated. As a result, neurotransmitter-filled vesicles move to the outer edge of the neuron. They quickly release their contents of neurotransmitters into the synaptic gap between neurons A and B.	The newly released neurotransmitters move through the synapse between the neurons. Eventually they bind to receptors on the nerve cell's surface. This binding is specific; different neurotransmitters bind to discrete receptors on the opposing cell.

Figure 13

SOUNDS CONFUSING?

The text uses a maritime metaphor to illustrate how neurons transfer information. In Dante's time, several ports faced each other across the Adriatic Sea, including the towns of Ancona and Zadar. In the analogy, a ship carrying a letter is sent to Zadar via boat; a response is then sent back via the returning boat to Ancona. These boats act like neurotransmitters, continually crossing and recrossing space to transfer information.

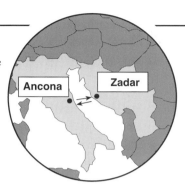

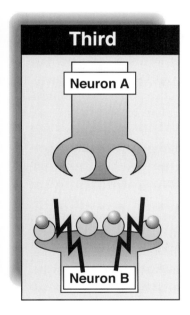

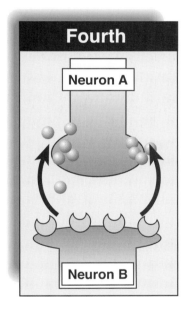

STEP 3: THE SIGNAL IS TRANSFERRED

The occupancy of the receptor results in the transfer of information to neuron B. The neuron becomes electrically stimulated, just as neuron A had before. The signal has been successfully transferred across the gap.

STEP 4: RECYCLING THE NEUROTRANSMITTER

Neuron B eventually releases the receptor-bound neurotransmitter. In turn, the neurotransmitter is recycled back to neuron A in a process known as reuptake. This reuptake is very important to our understanding of certain anxiety disorders.

Zadar, the supplier might choose a boat that regularly traveled between the two ports. He might write a note, saying something like "Here is a sample of our merchandise. If you like it, you can put in an order. Whether you order or not, please return the sample to me". The supplier in Ancona would put the package containing both the note and the sample on a ship bound for Zadar. After a period of time, the ship would arrive at a dock in the Serbian seaport, and the package would eventually reach its destination. The local merchant would examine the quality of the textile and become very excited at what he saw. Immediately, he would write a note back enclosing both the sample and his order. The ship would leave Zadar, sail across the Adriatic and eventually arrive back in Ancona. The supplier would retrieve his sample, happily read the response, eventually send over more textiles.

What does a story about medieval commerce have to do with the communication between two groups of neurons? It turns out that two neurons lying next to each other are physically separated by a watery space, just like the towns of Ancona and Zadar have the Adriatic between them. Scientists call this space a synapse. If the neurons are going to communicate with each other, they will have to send information across this synapse. Their solution is surprisingly similar to the way seaports like Ancona and Zadar communicated. The neurons use mobile intermediates called neurotransmitters to cross the watery distance to talk to each other. These ship-like neurotransmitters are actually small biochemicals. How they work is illustrated in Figure 13 and described below.

Let's say that neuron A becomes electrically stimulated and wants to communicate the news to its neighbor, neuron B. Neuron A does this by sending out a small flotilla of these neurotransmitters into the synapse. The biochemicals cross the distance and, like medieval merchant boats, dock at molecular "ports" on the other neuron. These ports are called receptors, the same kind of protein molecules we discussed in the chapter on lust. This docking is specific, even shape dependent. Just like a large ocean-going vessel could not dock at a tiny slip, specific neurotransmitters only dock at receptors with which they fit.

The docking of neurotransmitters to their individual receptors is a big deal. The neuron carrying the receptors can become quite stimulated, producing its own electrical signal, if enough ships bind. And when the neuron becomes activated from its resting state, that means the signal has been "transferred". This is exactly how signalling information is carried from one neuron to the next. By using neurotransmitters to cross the synapse, neurons can communicate with each other.

Sometimes the information neuron A wants to communicate to neuron B is inhibitory rather than stimulatory. For example, when neuron A is electrically activated, it may want to send a warning out to neuron B to ensure that it will not be stimulated in response. Neuron A still sends out its flotilla of aquatic neurotransmitters, and these still cross the synapse and bind in the usual way to the port-like receptors of neuron B. But rather than neuron B becoming excited by the newcomers, neuron B gets the message and shuts down. Quite appropriately, such interactions are said to be "inhibitory".

What then happens to these seafaring neurotransmitters once they have docked? They are not held prisoner, certainly. Receptor-bound neurotransmitters are eventually released from the receiving neuron after a period of time. Now untethered, they sail back over the synaptic ocean, eventually arriving back at the starting neuron in a process known as reuptake. This is just like the ship sailing back to Ancona after having docked at Zadar, carrying the textile sample back to its owner. Neurons care a great deal about the whereabouts of their neurotransmitters. As we shall see, imbalances in these sending and reuptake processes can lead to changes in behavior.

The genes of anxiety

With this information in mind, we are now ready to look under the hood of the neural cells involved in fear and see what makes them work. Valuable information about molecules and anxiety have come from two lines of research: genetic work with animals, and the effects of anti-anxiety drugs in emotionally disturbed humans. Though much has been learned in both the laboratory and the clinic, none of the efforts has yet isolated the "gene for human anxiety". We will examine briefly some of the animal work and then get to the work most relevant to human beings, the effects of certain medications on human behavior.

Many animal experiments have been done in an attempt to find genes governing anxiety. One of the most common modern genetic tools is to use what are called knock-out animals to assess behavior. What is a knock-out animal? Researchers have found a way to selectively and deliberately destroy a target gene in a developing mouse embryo. When the mouse grows up, it is genetically intact in every way except one: it is missing a single gene. The equivalent of a molecular smart bomb, these techniques allow the researcher to assess the function of the gene by seeing what happens when it is absent. Though roundabout in many ways, solid work has been done and a number of candidate genes have been identified that appear to play important roles in fear and anxiety.

Unfortunately, none of this knock-out work applies readily to humans. The normal perils associated with ascribing animal data to people are relevant here, of course. But there is another problem, one that we mentioned earlier in our attempt to understand the emotions we feel. Some categories of emotion do not reflect biological realities, but only human attempts to organize subjective experience. Is the same going to be true for the categories of fear such as phobias and OCD? The only way to find out is not to look at animals but to look at people, even though many brain mechanisms are conserved throughout evolutionary history.

Fortunately, we are not left with the genetic equivalent of broken altar pieces scattered to the four corners of the Earth in our attempts to find fear genes. Evidence is beginning to emerge that may actually unravel the complex interactions of molecules that govern human anxieties. Where do the data lie? – by looking at abnormal psychiatric behaviors and attempting to treat them. In many cases by accident, we have discovered medications that actually change fearful behaviors in people, many times relieving them of life-threatening symptoms. These medications have served something like a flashlight for the research community. By figuring out how the drugs fix the behaviors, we can work backwards to figure out how the behaviors are started in the first place. Described below are two such success stories, one related to general anxiety, the other to obsessive-compulsive behaviors.

Valium lessons

In the early 1960s a medication was introduced in the United States that became one of the most frequently prescribed drugs in history. Its chemical name is diazepam and it was found to have the extraordinary ability to reduce anxiety in the people that took it. You probably know diazepam better by its street name, Valium, a drug belonging to a class of molecules called benzodiazepines. By examining the mechanism of diazepam, scientists have learned a great deal about the molecular basis of fear and anxiety in the human brain. We will look at that mechanism as well, and hopefully gain similar insights. To discuss in a meaningful way how Valium works, there are two pieces of background information we will need to mention:

1. You may recall from our discussion of medieval commerce that when neurotransmitters land on a cell after their synaptic voyage, it is just like a ship sidling up to a dock in a port city. The dock, as you remember, is really a receptor on the surface of the receiving cell, and usually made of protein. The binding of the neurotransmitter to a receptor is not a neutral event to the cell carrying the receptor. The neuron may either become very excited or completely shut down its activity (the neuron may even become excited just long enough to tell still other neurons to which it is connected to shut down and stay that way). Either way, information is transferred. This shutting down aspect is important in our understanding of how Valium is thought to work.

2. You might be interested to know exactly how some of this shut down occurs after a neurotransmitter binds to its receptor. It turns out that many neurons, like a few chefs of whom I am aware, have a love affair with salt. There are gates on the surface of neurons that let certain components of salt (like chloride, for example) either in or out of a cell. The neural cells involved with Valium also have a pre-occupation with salt. In their resting state, they keep chloride out of the cell. While keeping chloride atoms in or out may not sound like a big deal, its movement into or out of a neuron can actually change the entire job description of the neural cell. These gates are fully capable of letting chloride in should the right signal come along.

How Valium works

With those two pieces of background information, we are ready to talk about how Valium works in us, and, indirectly, how anxiety may work in us too. First, we will talk about a receptor system.

In our brains, there are neurons which make a neurotransmitter known by the strange name of GABA (γ-aminobutyric acid). Like all neurotransmitters, this GABA is capable of floating across the synapse and binding to a protein receptor. Not surprisingly, the docking protein is called the GABA receptor, and many neurons (like those in the amygdala) carry it.

What happens when GABA binds to its receptor? The receptor acts something like a gatekeeper, controlling the movement of chloride across the cell's border. When GABA binds to the receptor, the following command is issued, "allow more chloride atoms into the neural cell, please".

And that's what happens. The chloride floodgates open and chloride comes rushing in. For reasons still not well understood, this influx actually creates an inhibitory signal, one felt by the entire neuron. The neuron, in essence, shuts down. The more chloride allowed in, the more this inhibition is seen. That's why its a big deal whether chloride is kept in or out of a neuron.

How does this gatekeeping relate to the normal experience of fear? When humans encounter a threat, just the right amount of GABA is released into the synaptic spaces between certain neurons in the brain. The chloride concentrations change, and certain inhibition signals are observed, just as we described in the previous paragraph. Eventually, the person feels fear. To get the feelings of anxiety, it is important that only the right amount of chloride gets into the neurons in areas like the amygdala.

With this complex biology in mind, we are ready to understand how Valium works, and, in so doing, obtain a molecular understanding of fear responses. Much like having two ships anchoring at the same dock, Valium has the ability to bind to GABA receptors, just like the neurotransmitter GABA. If Valium binds to the receptor first, an interesting command is issued. The command is "make sure that when GABA comes it binds to the dock more tightly and stays there longer". In the parlance of molecular biologists, Valium is said to "increase the affinity" of the neurotransmitter for its target receptor. Remember that a bound GABA molecule signals the neural cell to open its flood gates for chloride; increased affinity simply means that more chloride gets into the cell. This alteration is important, and not natural, but its effect is stunning. As a result of increasing the affinity of GABA for its receptor, Valium reduces the feelings of anxiety many people experience. It is truly an anti-anxiety medication.

What does that mean in the hunt for the genes controlling fear behavior? Nothing so far. Valium binds to GABA receptors wherever they exist in the brain, and they exist in many places, not just the regions involved in fear responses. Thus, taking Valium is like hitting the brain with a sledge-hammer. We know that pieces get broken, and that the breakage makes people relaxed. However, we don't know which pieces were the really important ones to break and which were simply by-products. Thus it is only safe to say that whatever anxiety-producing tissues do when a person becomes overly fearful, Valium instructs that tissue to do less of it when taken at sufficient dosage. Since the GABA receptor is itself a protein, meaning that it is encoded by a gene, a genetic link may one day be established between individual shapes of the protein and a person's susceptibility to anxiety attacks. As we have discussed, there is no gene for anxiety that has been isolated to date, and so at present nobody knows.

The picture is further muddied by the fact that other types of anxiety appear to operate by pathways that have very little to do with GABA. As discussed previously, the anxiety classified as OCD seems to work with different types of genes and other molecules. We know this because we have stumbled upon medications that can in some cases almost miraculously change the OCD behavior. Exploring how these work has given us new molecules to consider as we attempt to understand the genes of fear.

A problem in reuptake

Just what are these alternative pathways that seem to work with people suffering from OCDs? Recall from our commerce analogy that once neurotransmitters have docked at their receptors and transferred their information, they don't continue to hang around the port to become a polluting nuisance. Like the supplier asking for his sample to be returned, the neurotransmitters are eventually recycled back to the neuron of origin. This process is known as reuptake and is a common feature of many kinds of neural interactions.

One neurotransmitter that has received a great deal of attention over the years is a molecule called serotonin. Famous as a mediator of many behaviors, it is commonly recycled back to the neuron that made it once its job is finished. The reuptake is not automatic, however. Serotonin needs some help to get it back into the neural cell of origin, and the brain generously supplies the assistance in the form of a protein. This protein is called, logically enough, the serotonin transporter. It has the ability to bind to serotonin and assist in the process of reuptake.

What do these mechanisms have to do with OCD? A great deal, actually, all coming from the unexpected finding that certain medications truly stop the anxiety-driven repetitive behaviors so typical of these patients. The trick was to find out how the drugs work. Certain anti-OCD medications actually work by interacting with the serotonin transporter. This interaction interferes deeply with the reuptake process. Such disruptions in the normal recycling events alter the behavior. Since this mechanism is different from GABA's and its interesting inhibitions, we all of a sudden have at least two ways of looking at the genes that govern our fears.

Because serotonin transporters are proteins – meaning that they are encoded by DNA – it is natural for researchers to look to alterations in their genetic structure in an attempt to isolate OCD genes. The idea that the transporter plays a role in anxieties has been strengthened by the finding that some families plagued by OCD often carry mutations in their transporter genes. The findings are, unfortunately, not air-tight. Attempts to replicate these data have met with mixed success. At this stage in the game, there are still no fully isolated and characterized genes for the disorder, though many researchers are optimistic that the isolation is just around the corner.

And that's an adequate way of describing the state of the research. We know a lot about a few things, each new piece of data giving us further hints as to how far we've gone and how far we have to go. There are many issues related to fear that cannot be explained by the comparatively straightforward invocations of neurotransmitters and medications. I am of course talking about that netherworld between the tissues upon which we can experiment, like the brain, and the possible property that emerges from collective neural interactions, the mind. To see how our discussion of fear works in the chronicles of the controversy, we are ready to switch gear and talk about our next installment in the historical record of the mind/brain debate. From there we can return to our definition of consciousness to see how the concept of a fear subsystem illustrates the structure of human emotion.

Mind and brain part III

Fear is an important component of human survival, and many Ancients have taken pains to describe both its presence and its behavioral anatomy. In the first chapter of this book, we discovered that people in Dante's time borrowed deeply from the Greeks in their attempts to organize the origins and structure of such feelings. As medieval attitudes slowly gave way to the first stirrings of intellectual Renaissance, these notions began to shift. Many years after Dante's death, the famed Rene Descartes gave us dualism, the notion that humans have a mechanistic body and an indivisible will. Descartes also gave us a problem, however, setting the stage for a debate that still rages in some circles today. How could a physical body and an unphysical will interact to give us the varieties of human experience? Fear is the perfect example, for it is felt as an attitude (that would be the mind), yet there are many physical manifestations, such as shaking and nausea.

There were two ways philosophers and scientists tried to solve the problem of a physical body and a nonphysical mind. The person representing the first method was Gottfried von Leibniz (1646–1716), who attempted to solve the problem by avoiding it. Leibniz declared that the physical body and the ethereal mind worked together like an unsuccessful predator stalking an elusive prey. They followed parallel courses, but they never actually met. Thus, it only *appeared* as if you got a headache when the robber bonked you on the head. You might be really shaking and quaking and needing to go to the bathroom, but you only *appeared* to feel real fear in a life-threatening situation. Because the two worlds did not touch, Leibniz reasoned that a physical experience had to be coordinated with an associated mental one from an external source. Who could do such a philosophical lip-synch for all of humankind? That was the province of God, said Leibniz. Deity alone had the ability to keep the two in synch. The theory is still known today as psychophysical parallelism, and, as you might expect, does not have a lot of modern research support.

The second solution was championed by philosophers such as George Berkeley (1685–1753). He created the notion of what later began to be termed idealism. This idea also attempted to avoid the problem of mind/brain, but by going in the other direction: Berkeley said there was no substance beyond that which could be carried in a thought. The physical world was nothing but groups of perceptions taking place in our minds. In this view, it was foolish to think that a bear seeking to attack us was any more real than any of the other countless virtual objects tucked inside our imaginations. Berkeley's thoughts in some ways paralleled Leibniz's, however, because he had to solve the problem about how an unmolesting bear might continue to exist if no one was perceiving it. Once again, the Almighty was invoked. Since God perceived everything all the time, His existence gave the objects both meaning and presence to the physical world, no humans required.

Another ingredient for consciousness

Both Leibniz and Berkeley have to appeal to the notion of Divine Memory and Awareness to demonstrate bodily interactions with worldly items. With Leibniz, God keeps everybody's perceptions on track with their physical experiences, inferring the presence of enormous cosmic listening and storage devices. Berkeley's world could evidence the presence of sound as a tree falling in the forest *sans* human because of the same omnipresent awareness and storage. This idea of storage device – memory – being linked to the idea of awareness is not foreign to us either, though we are talking more about neural interactions than hermeneutical ones. In Chapter Two, we discussed the role of long-term memory in consciousness; in Chapter

Three, we talked about short-term buffers. We are now ready to introduce the third ingredient in our discussion of human consciousness, one that serves as a liaison between our long- and short-term skills. Called working memory, this notion is deeply involved in our experience of fear. Even if described in divine terms, the notion would not be foreign to either Leibniz or Berkeley.

The idea of working memory

Anyone who has undergone a terrifying, life-threatening experience and lived to tell the tale will say one thing: the fear is unforgettable. Pedro certainly never forgot his perceived brush with death as a stowaway on board the ship to New York. Fear and memory are inextricably linked, a phenomenon true even at the anatomical level. We have already discussed two types, explicit and implicit memories.

This linkage means that understanding how memory works becomes an important component in knowing about fear. It is also important for another reason. As you recall, we have been detailing that various kinds of memory are important ingredients to the notion of consciousness. In this chapter we will talk about a notion termed working memory.

The concept of working memory can appear fairly complicated, and, even after years of argument, controversial in some circles. To understand this complexity, we need another medieval analogy. This one is taken from a comment Dante made about a very famous French dynasty, one famed for its greed and its cruelty. As you recall, the Capetian dynasty was named after its founder Hugh Capet. One particularly notorious member, one with whom Dante had a near-personal encounter in real life, was a man named Charles of Valois. Here's what Dante had to say about him.

> I see a him, not far off, that brings forth
> Another Charles from France. It shall make clear
> To many what both he and his are worth.
> He comes alone, unarmed but for the lance
> Of Judas, which he drives so hard he bursts
> The guts of Florence with the blow he plants.

Charles of Valois was one member in a line of Capetian kings whose history was intertwined with the poet's, and Dante is actually talking about a historical event to which he was witness. As is so true for much of human history, this story revolves around money, intrigue and, most of all, human greed.

Florence was increasingly becoming an independent center of commerce, under the wily administration of several city councilman (Dante just happened to be one of these crafty city fathers). It was of course dangerous to allow an important trade and rich banking center like Florence to have too much autonomy in unstable Northern Italy. Recognizing this fact, Boniface VIII, the not-so-local Pope, conspired with an even more wily character named Corso Donati to capture Florence and bring it to heal. Donati was a bit too weak military-wise to do it himself, not being an imperial King and all, so the Pope looked around for a substantial ally. And who was around? No less than one of those greedy Capetian kings, Charles of Valois. He was persuaded to march on Florence ostensibly to re-establish "peace and order".

But that's not what happened. No sooner did Charles come to Florence than Donati also entered the city gates, armed with a band of thugs. Immediately he and his ilk set to work disemboweling the town; houses were burned, nobles were killed, rich warehouses plundered, Dante himself was condemned and banished. In gratitude for his "help", Charles was paid a huge bribe, smiled happily and took his army south, having handed over the city for his good friend Boniface. Such betrayal was the reason Dante used the metaphor "Judas" to describe this powerful Capetian king.

Why do I bring up this bit of history? As we read through this section of *The Purgatorio*, we realize that Dante is describing events that nearly cost him his life. His non-neutrality regarding the personal character of Hugh Capet's line is thus understandable. As he wrote these lines, it is quite possible that he began to re-experience his fear. Understanding what his body experienced as he scribbled the events of the Fifth Cornice might give us hints to the biology underlying any fear. Here's what he probably experienced

1. Elevation in pulse.
2. Changes in blood pressure.
3. Hair standing on the back of his neck.
4. Shaking.
5. The need to evacuate both bowel and bladder.
6. Immobility – "freezing" – like a deer caught in modern-day head lights.
7. Jumpiness, even to the point of severe startling.

All this physiology stems from a memory. Many researchers call such reactions the fight-or-flight response.

As we've discussed in the last two chapters, Dante could not have written these words without having access to several memory apparati within his brain. He could not have recalled the brutal scenes of the invasion of his home town, an event that happened six years before he wrote *The Divine Comedy*. But how did those memories get to be long-term in the first place? In the last chapter, we mentioned the role of short-term buffers. However, we also said that both long-term and short-term devices need to talk to each other for proper processing. What liaison is responsible for such important communication? And what does all this have to do with consciousness, or even the emotion of fear?

The answer to these questions turns out to be involved in a third memory system embedded in our brains, this cognitive talent of working memory. Working memory is such an important concept that many researchers have built entire theories of consciousness around it. It is best to define the concept by giving an example. Go find a blank piece of white paper, one that has no markings on it, and place it on a flat space near this book. With the paper close by, remember, if you will, the following list of Capetian kings: Hugh, Robert, Philip, Charles, Louis. Now look at the blank piece of paper and (1) repeat the names you just saw and then (2) recite what you had for dinner in the previous six days.

Now, can you repeat the list of Capetian kings you saw, and in their proper order, before you identified your dinner? Use your piece of paper to write down their names. Can you do it? If you are having trouble, don't worry; there is actually a better than evens chance that you cannot name all of them. The reason for that is straightforward. Deep in your brain you possess a workspace – much like a desk in an office – that allows you to memorize things. You put those five names of kings into that space when you sought to memorize the list. The dimensions of the desktop are limited, however immediate though the memorization seems, which means the workspace has a finite capacity. Because the space is limited, something will have to go when you try to put new things in. That's what happened when you tried to remember your dinners. You had to bump some of the French Kings off the desktop, and that's why you might not have been able to remember them.

This workspace is the heart and soul of the concept of working memory. Many years ago, working memory was called short-term memory. It was eventually found that not only were short-term events stored there, but that active mental processing occurred there as well. To distinguish it from the buffers, it is called working memory. There is room for several pieces of information to be held in working memory at once, hence your initial ability to remember some of the Capetian Kings. There is also room to allow variables to be compared and contrasted, hence your knowledge of the meals you ate.

Working memory is now known to be involved in nearly every active thinking process in which human beings are aware (including the processing of fear). That's why so many people think working memory is an integral part of consciousness. It even serves as a liaison between the other two types of memory we have discussed, long-term storage and short-term buffers. How can that be? To gain one last insight, let's take another example.

You and I both can hold the idea of a kingly crown in our working memories. It got there from a cooperative interaction with the other two memories. Your short-term buffer retains the memory that you just read the word "crown" in the previous sentence; it may even help you form a visual representation of the concept. The idea of "crown" was retrieved from some long-term memory device, perhaps from a short story you read as a child. Thus, working memory used both the buffers and the long-term talents. That's why scientists consider it a liaison. In order for your working memory to create meaning from what is being processed, it requires input/output channels to both your short-term buffers and long-term memory. It thus acts something like a broker, allowing both distant and recent events to be integrated into an awareness of present reality. This awareness function is why the concept of working memory is included in our definition of consciousness. Memory systems are thus not only important in the generation of fear, but in the generation of our awareness of it (see Figure 14 for a summary of interactions).

The rest of the story

We still have a long way to go before we truly understand how phenomena such as working memory interact with phenomena like serotonin transporters in the generation of human fears. The same can be said for GABA, chloride gatekeepers and a hundred other molecules we don't have space to mention in this chapter (indeed, entire books have been written on the subject of fear, tissues and anxiety). The upshot is that real progress is being made, and with every data point uncovered, we seem to understand both more and less how fear is generated.

How working memory works

Acting like a desktop with limited workspace, working memory holds and processes inputs from a variety of sources. Here's how it works.

Working memory serves as a liaison between two types of inputs, short-term buffers and long-term memories. A variety of sensory data (such as visual signals, auditory inputs, even data from language systems) can be stored simultaneously in short-term buffers, which usually interact with the working memory space one at a time. Working memory also communicates to and processes information from long-term storage memories. As shown below, working memory helps integrate the information stored in the buffers with the long-term memories.

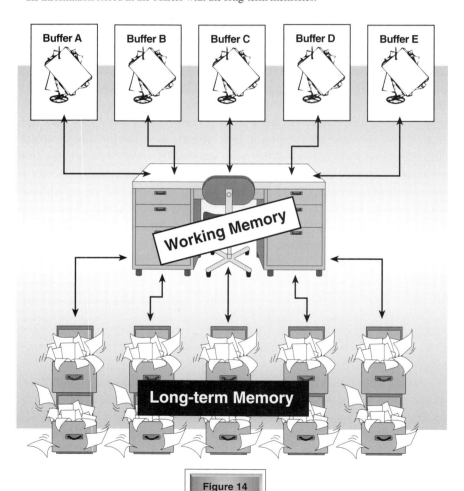

Figure 14

As we learn more about the roles of various genes, tissues, brains and minds, it becomes increasingly clear how inadequate our human categorizations of emotion can be. This chapter is a prime example. Even though we may have conscious feelings of greed, we have yet to isolate a region of the brain exclusively devoted to avarice. Rather, we have to look to psychologists to give us hints as to some underlying process, and then hope that we can use that process to go looking for a neuron. In this chapter we looked to fear in hopes of illuminating something about greed. A fair argument can be made that this is a rational thing to do, but not because we hope to isolate a greed-generating gene at the end of our search. The presence of an emotional category does not *a priori* mean that there is a concomitant brain function devoted exclusively to it.

We have encountered this kind of thing before. As we move through our history of mind/brain problems, for example, it becomes clear that human beings have a habit of over-extending their interpretations from scientific observation. Historically, as now, such sloppiness can create border skirmishes with subjects science could never hope to address (one could never publish Leibniz's theory that God created a parallel mind in the journal *Nature*, for example). We do better to train our insights into understanding how things like working memory might relate to the emotions of fear.

The idea of consciousness brings up our final subject before we move downward to the next cornice. As you recall, our model of the composition of emotions postulates an interaction of awareness with some emotional subsystem; we have been attempting to isolate both the ingredients of the subsystem and the ingredients of consciousness in our quest to evaluate the postulate.

A parting thought

From our chapter on lust, we learned that emotions are usually things that happen to us, rather than things we will to occur, and that many parts of our emotional responses lie beneath the surface of our consciousness. This unawareness is so complete that some scientists have begun to doubt the existence of emotions as a legitimate brain function at all. Rather, reminiscent of our discussion above, they may simply represent human attempts to categorize subjective experiences. From our chapter on appetite we learned that emotional subsystems may be shared up and down the evolutionary chain. The insight is that these emotions exist in order to ensure the survival of the species to the next generation, in this case, causing hunger so that the creature possessing the appetite keeps up its fuel supplies. This makes animal experiments both very useful to do and, as we discussed in the chapter, dangerous to interpret. In our present chapter on avarice, we are attempting to show that emotions, if they exist at all, may be easily mislabeled. There may be no such thing as greed, even if there is a powerful subsystem that governs our fears. As we shall see in a later chapter, greed may even be part of our drives for aggression.

Taken together, we seem to be defining emotions the same way uninformed people might try to find Florence on a map: determining its location by finding first where it is not. As confusing as it may seem, we must understand there is a simple underlying survival motive for keeping emotions around – even if they seem scattered to the four corners of our brains. This need not be totally uncomfortable, for in many ways we experience such unexpected unity on a daily basis. Take those five altar piece panels painted by Giotto and blown to museums all over the planet. As I gazed at *Pentecost* in the London gallery, it seemed a concluded, solitary art piece. I was in the good company of certain art critics, and it was indeed mislabeled as a single work for many years. They did not realize, without a little digging, that London had only one-fifth of the original design. Considering both the role they play in survival and how little we know about them, that's not a bad way to think about human emotions too.

CHAPTER FIVE

Sloth

"He seized the witch, and with one rip laid bare
All of her front, her loins and her foul belly;
I woke sick with the stench that rose from there.

I turned then, and my Virgil said to me;
'I have called at least three times now. Rise and come
And let us find your entrance.'"

-*Canto XIX, The Purgatorio*

"What if you were sleeping, and the person next to you started to take their clothes off?" The professor addressed the class in a soft voice, in stark contrast to his noisy charges, shifting restlessly in their seats. They were instantly silent. "Do I have your attention?" he smiled, now clearly heard. "It gets worse. What if they were so filthy that when they took their clothes off, the smell literally woke you up, and gave you a horrible fright?" A few members of the class wrinkled their faces, and several looked a bit puzzled. "It would get my attention too," he confessed, "But that's exactly how Dante ends his journey into the Fourth Cornice – the level devoted to those guilty of the great sin of sloth."

Now most of the class had puzzled looks on their faces, including myself. Sloth? Clothes? Doesn't sloth mean *lazy*? The professor was talking about a truly coarse series of lines in Canto XIX, near the end of the journey into the Fourth Cornice. But he also seemed to read our minds, and continued his lecture this way. "A lot of people think that sloth means indolence, or laziness or being eternally sleepy and inattentive. And that's true, to some extent. In fact, if the chapter closes with this get-your-attention smelly striptease, it opens in just the opposite way, with poor Dante barely awake. In fact, there is a suggestion that the level is enchanted, and those that stay there get drowsier and drowsier."

"But Dante did not sleep for long," the professor continued. "He is awakened by what appears to be a herd of noisy animals, running around the cornice with a great deal of shouting, energy and zeal. The creatures did not turn out to be animals. They were simply a large group of extremely hyperactive human beings."

"Why were they running?" the red-haired boy's hand shot upwards, even as he asked his question. I noticed that he looked rather disheveled this morning, as if he had pulled an all-nighter the day before.

"The reason for their activity gives us our definition of sloth," the professor explained. "It isn't this dreary, sleepy state of torpor we think of today. Dante said that the people condemned to this cornice were those who recognized The Good, but were too lazy to pursue it. Their punishment was to become all hurry and zeal, running endlessly around the cornice, even as they were slack in life. They are so active that they do not even slow down to talk to their visitors. And that's a first for the journey of the poets in *The Purgatorio*."

It wasn't a silent place, the professor explained to us. The usual kinds of chants and choruses could be heard as the inmates practised their medieval exercises. Two ran ahead of the others shouting something about the diligence of Mary and Caesar, two ran behind the others shouting something about Israelites and Romans. After the dust settled, Dante dozed off again, perhaps unable to resist the sleepy enchantment of the place.

"Just before morning, Dante had a strange and rather repulsive dream," the professor continued in the class. "He began dreaming of The Siren, a woman who lures the souls of men to temptation the more they contemplate her visage." Virgil somehow became aware of the contents of Dante's dream and was suddenly alarmed. He apparently considered it dangerous to be thinking of sirens and so Virgil miraculously entered Dante's reverie and stripped The Siren bare.

"When he did this, he exposed her filthy body. A stench so great arose after this sudden act that Dante was immediately awakened," the professor explained. "And that's when they leave this contradictory place, filled with enchanting sleep, rude awakenings and eternal hyperactivity."

What this chapter is about

In the odd world of *The Purgatorio*, Dante relates that Virgil's awareness of The Siren wasn't all that mysterious; Virgil was alerted to her occupancy in the dream by someone Dante calls The Heavenly Lady. The opening quote in this chapter relates what the Roman does next to expose her sinfulness. The most interesting aspect of this level of *The Purgatorio* concerns what can be seen in the detail of this incident: the familiar presence of both sleeping states and waking states in human beings. Due to sloth, the inmates are condemned to run in a world of continuous awareness, headlong and helter-skelter. But the punishment for sloth lulls any visitors to the cornice to sleep. Dante succumbs to it twice, once at the beginning of the visit to the Fourth Cornice and once near the end.

The questions biologists could pose to Dante might be about his tendency to nod off, and equally what keeps the inmates wide awake. What is the biology behind sleeping? What is the biology behind awakening? What are the rhythms that appear to govern these alternating cycles of sleeping and awakening? Most of us can relate to the inmate's activity because we certainly spend a great deal of our lives in the awake state. And we can relate to Dante because we also spend a great deal of time, about a third of our lives, sleeping.

In this chapter we are going to explore the biology behind a familiar cycle, the one that puts us into our beds at night and gets us up in the morning: we are going to examine the sleeping and awakening of human beings. We will start by looking at some of the organs, tissues and genes that govern our cycles of dozing and waking. We will end our discussion by seeing how this biology contributes to our understanding of minds and brains, even consciousness. Along the way, we will once again observe the invaluable contributions animals have made to our comprehension of human behavior.

We will also observe an interesting characteristic, a first in our exploration of The Genetic Inferno. While the study of these pacemakers constitutes a human behavior, the time-keeping mechanism itself is not necessarily an emotion. As we shall see, disruptions in the pacemaking functions can generate plenty of emotional behavior. But by themselves, the tissues that generate our internal clocks are not necessarily the source of moods. As in our discussions of lust, appetite and greed, we are looking at the biology necessary for the formation of these feelings, and leaving it to our model of emotion-system-plus-consciousness to show us where the feeling may lie. While there may be an emotion of "tiredness", there is also a distinct biological need to sleep. And it is to this biology that I would like to turn. In many ways, it is the clearest road we have yet trod that connects specific genes to specific behaviors.

With those thoughts in mind, let's start talking. We will begin by trying to find a definition of something scientists like to call circadian rhythms.

The biology

We have all heard about experiments where the sleep cycles of human beings are measured in the cruelest of ways: some poor souls are thrown into a lightless cave for a period of time and their various activities observed and recorded. Every time this experiment is done the researchers find the same astonishing thing. Even in unlit living quarters, activities like sleeping and waking, physical exertion and hunger, even reproductive fecundity, fall into a steady carousel of events that turn every twenty-four hours or so. Our physical body can detect the length of a day somehow, even if there are no external cues to coach it along! Working something like a quartz crystal inside a well-made Swiss watch, scientists believe humans possess some kind of awareness/pacemaker, whose ticking comes from within the brain. This internal ticking is so loud in our body's chemical "ears" that even re-exposure to daylight and night-time only corrects the clock by an hour or so.

Some general comments

It isn't just humans that respond to daily cycles of activities. Most creatures that have ever been studied display daily oscillations of some kind, the general definition of circadian rhythm. This commonality is fortunate because it has allowed us to study "simpler" organisms, isolate their genes and see what relevance the work has towards people. The kinds of creatures that have yielded data have been enormously diverse; in this chapter alone we will talk about how systems ticking away in hamsters, fungi, fruit flies, frogs, and mice have made fundamental contributions to our understanding of human circadian cycles.

Even plants get in on the act. In fact, for many years, plants were thought to be the only creatures that exhibited such oscillations. Anecdotally, of course, it was known for centuries that the life cycle of botanical organisms was deeply enmeshed and responsive to external environmental cues. Formal descriptions started with a French scientist (an astronomer named d'Ortous de Mairan), who noticed that leaves of certain plants opened and closed in a daily fashion. In 1829, Augustin de Candolle demonstrated that this closing occurred on a rhythmic schedule even if the plants were kept in total darkness. These two important pieces of data showed that there was a periodicity to the behavior of these creatures, and that this periodicity was independent of light. A pacemaker appeared to exist *internally*, driving the leaves to open and close using what appeared at the time to be mysterious interior commands.

Standing on this initial work, the tools of the modern biologist began to unravel the pulsing biorhythms of creatures other than plants, such as brain-carrying mammals. It was found that the "ticking" is actually composed of a series of rhythmic firings of individual neurons. As you recall, this firing is termed "depolarization", and is meant to signify that the neuron has received some kind of stimulation. The automaticity of circadian nerve cells simply means that the neurons are self-stimulating on a clock-like schedule. As we shall see shortly, individual neurons possess distinct firing patterns, and the sum total of many neurons in the brain constitutes the pacemaking functions of the creature's internal clock.

With such individually complex neural contributions, is it possible to create an overall model that explains circadian rhythm in animals as diverse as fruit flies and frogs? An overarching conceptual template has indeed been proposed on the basis of such knowledge, one that appears to involve three generalized structures. These structures have been observed in most circadian-responsive creatures.

1. *An internal pacemaker.* The creature must possess some kind of internal rhythm-generating device. This device needs to have a couple of characteristics that at first sight might seem contradictory: (1) it needs to be able to operate independently of the environment; and (2) it needs to be able to respond to cues from the environment and make adjustments according to what is sensed.

2. *Linkage to the environment.* Some kind of mechanism must exist that provides the cues for the adjustment mentioned above. In other words, there needs to be a link between the external environment and the creature's internal cycle. Acting something like a smart antenna, this input device must be capable of both detecting and discriminating between alternating light and dark patterns in the great outdoors.

3. *Linkage to the creature.* The organism's pacemaker must be able to create outputs based on what it senses, outputs that can be understood by the rest of the body. This is of course important for sleep/wake cycles, but is also critical for a surprising number of other functions such as hormone secretions, breeding cycles, even physical activity. All of these processes need to be linked in some fashion to the timing device in order to be experienced in a circadian fashion.

Taken together, these three components constitute such an incredibly complex set of functions that discerning how it all works together might seem an impossible task. While parts of the story in humans are convoluted, enough research has been done that a clear picture seems to be emerging. Some of it is very surprising. Many researchers first thought that the human clock was a single squirming cauldron of neural complexity, made of widely dispersed tissues involved in subtle, highly organized, mostly inaccessible communication networks. Once the pacemaker region of the human brain was found, however, it was discovered that the clock was much less federal. *Every* cell in the clock area appeared to be governed by its own independent pacemaker. The sum total of these insular ticking devices, each possessing their own oscillation pattern, gave us the human circadian rhythm. This meant that studying the clock-speed of an individual cell would be as important as studying the clock-speed of cellular groups. And that's where we will be headed as well. We shall begin our exploration by first looking at the clock region of the brain (you will be surprised at how familiar the region we'll be exploring is), and then we shall move to contributions of individual cells, describing in brief what is meant by clock-speed.

An internal pacemaker

The clock in the human brain is found in the anterior hypothalamus. You did not read that wrong. This anterior hypothalamus is the *same* place that some of the other emotions (such as lust and hunger) find their neural workshops. The regions within the hypothalamus that harbor the clock have been relegated to a subdivision known as the suprachiasmatic nucleus (we call it the SCN for short). In humans, the SCN is composed of small paired structures, each containing about ten thousand neurons.

How is it that thousands of tiny neurons can work together to give humans the ability to keep track of time internally? To understand the principle behind how the SCN works, I would like to introduce you to a professor friend of mine named Harry. He's not a biologist at all, but a historian, and one with a rather odd flair for physics. When I told him I was writing this book, he invited me to his musty old office for a chat about medieval clocks. We ended up meeting three separate times together, the first of which involved medieval timepieces. His explanation about clocks makes for an illuminating analogy about the individual contributions of neurons to the SCN.

Harry's office

"Want to see a picture of an old medieval clock?" he asked when I sat down, his eyes twinkling, the smell of cigarette smoke in the room. He took out an old picture book. "This is a drawing of the clock found in Rouen," he gestured, referring to the city in France, and put his own aging fingers on the fine detailing of what turned out to be a schematic. "Pretty thing for being so complicated, isn't it?" he grunted.

I agreed. That complex detailing is *the* thing that strikes you about this mechanical device, even if it was made in the fourteenth century. My friend told me that mechanical clocks were perfected in Dante's time. The great breakthrough, he explained, was learning how to drop a weight in an organized and predictable fashion. "You know better than I that all machines need a source of energy to run. So do clocks, like that coming from a wound-up spring, or an alternating current, or even a falling weight, which is what people in Dante's time used." Harry then went on to explain that the falling weight turned a wheel, which in turn moved a system of gears, which turned the hands of the clock. The trick was to carefully regulate the velocity of the falling weight, making sure that its descent was arrested at regular intervals. That arresting mechanism was called an escapement. An escapement allowed the teeth of the gears to slip, one at a time. The secret was in designing the proper kind of escapement. And that's what the fourteenth century gave us, the proper design for an escapement.

Designing one turned out not to be an easy task. Harry explained that a lot of trial and error went into the effort before the right one was made. The most successful designs employed a bar-shaped thing Harry called a foliot, "It was really an oscillating horizontal rod that was physically attached to a verge . . .", which was a vertical spindle, ". . . which in turn had two protrusions which we still call pallets, and when the pallet meshed with a gear that was already driven by a weight, it stopped the wheel it was attached to for just a moment. That gave the folks in Dante's time the needed interruption. Do you understand?" I nodded my head as if I did. He grunted and put the book away. We looked at a few more pictures of fourteenth century clocks, one in Salisbury, the other in Wells, England, and saw a similar complexity.

I didn't really understand the mechanism, but that was hardly the point then, nor the reason for bringing it up here. The impression I had after viewing the Rouen schematic was one of great complexity for a rather straightforward, singular task. Even in the fourteenth century, there were so *many* different devices in the clock, moving with different rhythms – wheels, gears, horizontal bars, spindles – all fueled by a dropping weight. Just the escapement physics alone was enough to show the ingenuity of what is usually called the Dark Ages. Yet these many devices all had to work together to accomplish a single goal, which was to tell medieval people what time it was.

This idea of complexity and unity of design is the very impression with which I wish to leave you as we behold the biology of circadian rhythms. Just like the gears and meshes in a medieval clock, the various neurons that make-up the SCN have their own individual clock-speeds, in this case, pulsing rhythms. Yet it is the sum total of these neurons that work together like an aggregate escapement to give our brains the twenty-four-hour pulses that are the hallmark of circadian information. The time-piece embedded in our brains is the collective effort of many different timing activities of individual neurons, just like the clock at Rouen.

This convenient *e pluribus unum* gives scientists an important intellectual scaffolding for framing the kinds of questions needing answering before we can find and characterize the circadian pacemaker. The first set of questions is the most obvious. How did we know that the SCN, with all its neural foliots, gears, and spindles, contained circadian rhythms? Studies of induced brain lesions in mice, knife cuts in frogs, even transplantation work in hamsters have shown that both the pacemaker abilities of these fascinating groups of neurons and their independent oscillating functions exist in the SCN. If one physically destroys these pacemaker neurons in a hamster, for example, all circadian rhythms are lost. If one then transplants SCN tissue from some donor to the damaged recipient, the rhythms are re-established.

This functionality has been demonstrated in a striking way, using a hamster carrying a spontaneous mutation scientists dubbed tau. This mutant animal possesses an abnormal circadian rhythm, due to the presence of a mutated gene. If the animal possesses one copy of the gene, its circadian rhythm (measured by locomotor activity) is shortened to twenty-two hours. If it possesses two copies, the periodicity is cut to twenty hours. When the SCN from a mutant tau animal is transplanted into the brains of a normal hamster whose SCN has been destroyed, the restored animal carries the attenuated rhythmic characteristics of the donor. Almost a quarter of a century of work has been done that points to a single result: the SCN is *the* neural hotspot for mammalian oscillators.

Taking hints from the animals, researchers began looking at the SCN of human beings in an attempt to compare its functions. Sure enough, our SCN was demonstrated to contain the human clock as well. Its characterization has been so thoroughly studied that therapies have been successfully employed that can "correct" its function when it is externally perturbed (as during jet lag, for example).

As scientists better understood the role of the SCN in circadian rhythms, it became clear that, in some creatures, the SCN was not the only pacemaker. Further research involving some of those hamsters we just discussed reveals an interesting finding. Buried deep within their retinas are neurons that appear to have their own circadian oscillations. Like their brainy SCN cousins, they too can generate their own rhythmic firing patterns, even if the cells are kept in complete darkness for long periods of time. The fruit fly appears to have a pacemaker embedded in the tiny structures that work as kidneys. Indeed, the fruit fly has an abundance of tissues that possess their own independent clocks (though it is not known if they work together in a coordinated fashion).

This individuality is a hallmark of all cells that possess circadian rhythms. That's also true with humans: individual brain cells of our own SCN contain their own individual timing sequences, as we shall see shortly. But there is a limit to their independence. The therapies described to correct sleep cycle problems have demonstrated to scientists that these individual clocks are both independent of and affected by the various sensory organs that allow light/dark cycles to be experienced. The second component of our model of circadian devices, as you recall, are these very links. In humans there appear to be two overall external linkage systems, and it is to these that we now turn.

Linkage to the environment

The first link to the outside world is the most obvious. Since timing cues revolve around visual perceptions of day and night, it is natural to think about SCN associations with tissues involving the visual system. One might predict, for example, that the SCN has connections that eventually couple its pacemaking neurons to the cells of the human retina. And that's exactly what one finds. There are both direct and indirect connections from the cells of the retina to the SCN. Indeed, the SCN's very location gives away some of its connective secrets: the SCN lies on top of a great visual switchboard, known to science as the optic chiasma, a structure that combines visual information from both eyes.

Is the visual system the only linkage the SCN has with the rest of the twenty-four-hour world? In one of the oddest pieces of scientific data to come around in a long time, the answer appears to be "no". Other regions of the body, not connected to our eyes, also appear to provide circadian information to the SCN. One of the most sensitive areas is called the popliteal region and exists – I doubt you will believe this at first – in the back of the human knee. It appears that the SCN can derive information about day and night from light that hits the hinter regions of our legs! This limb mechanism has only recently been discovered, and provides some novel hypotheses about circadian signaling, invoking not the presence of neural information (like that from the retina) but signals sent via the bloodstream. It is hypothesized that light hitting the back of the knee may cause some kind of signaling change in our blood. This signal then hitches a ride on the vascular train until it arrives safely at the SCN, where it informs the brain of day/light changes.

Linkage to the creature

In humans, it thus appears that we have both neural and blood-borne signals capable of telling the brain about the light and dark of the outside world. Following the next step in this story obviously means finding out what the brain does with the information once it has been received. And that leads us to our final set of tissues: the third set of structures in our circadian model concerns reactive output.

After the SCN has received information from various sensors concerning days and nights, its neurons issue a series of biochemical instructions. One type of instruction is a hormone that has received a lot of enthusiastic press (some would say *overly* enthusiastic press) lately. The hormone is called melatonin (see Figure 15). When darkness is encountered, the SCN sends a series of signals to specific regions of the brain whose end result is the secretion of melatonin. Once in the bloodstream, this hormone helps mediate several cues involved in human sleep, from administering the signals that inaugurate drowsiness to depressing body temperature. When light is re-encountered, the SCN sends out a series of signals telling those same brain regions to stop secreting melatonin. The end result is a drop in the hormone levels throughout the body. Many researchers believe these inhibitory signals help arouse the sleeping person to wakefulness. The whole thing is thought to work something like a sleep thermostat, continually adjusting itself to the changing characteristics of the outer world.

Melatonin is hardly the only story in the output side of circadian biology. There are many different kinds of output devices besides hormone secretion that are intimately linked to the pacemaker regions of the brain. Indeed, the input signals are so obvious – and the output effects are so clear – it is tempting to oversimplify the picture by invoking the SCN as *the* central clearing house for processing circadian signals. As we have seen in other creatures, however, other tissues besides the SCN also possess pacemaking functions, making this simplified comfortable picture an illusion. The information about modulation of rhythms through the back of a human knees should serve as a warning signal that the picture is much more complex in us too.

Oversimplified or not, the finding that the SCN was vital to circadian rhythms meant that we could study the SCN's component parts and learn important information about how we react to our environment. And that's exactly what researchers did. As discussed earlier, it was found that the SCN is not a monolithic pulsating organ, but is composed of individual cells, each possessing their own clock-speed. Each neuron has its own specific, independent, firing patterns.

How did researchers determine that such differences exist? It is possible to isolate neural tissue from the SCN and place the neurons into specialized dishes. Remarkably, the isolated cells continue their individual firing

Melatonin and the sleep cycle

The hormone melatonin is deeply involved in the regulation of circadian rhythms in human beings. Here's a brief summary of how it works.

THE NECESSITY OF LIGHT

Circadian rhythms can be affected by the presence or absence of light (either through the eyes or skin). This regulation is mediated via communication with an area of the brain known as the suprachiasmatic nucleus, or SCN. The panels below show the areas of the brain involved in the circadian rhythm, and what occurs (1) when light is present, and (2) when light is absent.

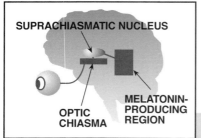

SUPRACHIASMATIC NUCLEUS

OPTIC CHIASMA

MELATONIN-PRODUCING REGION

Neurons, via the optic chiasma, connect the eyes to the SCN. The SCN is also associated with brain regions responsible for the manufacture of melatonin, making it an ideal "clearing house" for circadian information transfer.

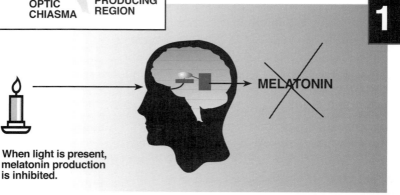

1

MELATONIN

When light is present, melatonin production is inhibited.

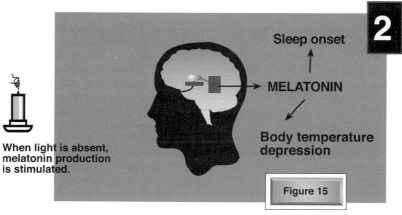

2

Sleep onset

MELATONIN

Body temperature depression

When light is absent, melatonin production is stimulated.

Figure 15

patterns even though they are removed from the brain. These patterns can be observed and recorded because the specialized dishes were "wiretapped" (using microelectrode arrays), allowing the scientists to listen in on the neural chatter. That's how their individual functioning was discovered. The neurons do not fire together in single bursts, like some coordinated capella choir singing in unison. Rather, the neurons exhibit a wide variety of rhythms, firing in different phases and for different lengths of time. Because these individual patterns are self-governing and can last for weeks, it has become obvious that cells can support their own independent clocks. And *that* means that exploring the inner workings of a single neural nucleus would provide important clues to the overall nature of circadian rhythms. In other words, it was logical to begin looking for some genes. That's exactly what researchers decided to do, and next we'll explore the fruits of their labor.

Our strategy for looking at genes

As we discussed previously, a great deal of knowledge has been uncovered by using animal systems as experimental models, and then discovering whether what is found in animals applies to human beings. Because so many creatures exhibit circadian rhythms, researchers were afforded a diverse variety of organisms upon which to build their models. True genetic experiments could be done with bread molds or rats and meaningful results obtained. To show how this works, we will explore what genetic mechanisms have been discovered in three different creatures: a fungus, an insect, and a mouse. We will then see what of this information, if any, applies to humans.

As we have experienced in previous chapters, understanding the world of gene regulation can be a fairly stout task. We will need to delve a little deeper into this world to make sense of some of the genetic processes in circadian rhythms. To talk about these mechanisms, I would like now to tell you about the second meeting I had with my professor friend Harry, the medieval historian with the interesting flair for physics. This time I took my one-year-old son to his office, an office that my son mistook for his own personal playground. I had yet another question for Harry about a medieval technological advance.

The second visit

Joshua was full of curiosity about Harry's office, for it is the kind of space filled with delightful looking gadgets, plants in desperate need of water, as well as dusty old and bright shiny new books of every description. Joshua's little hands seemed to have an especial affinity for the professor's portable photocopying machine. The little guy traced his stubby fingers around the outline of the on/off button, and as soon as Joshua pressed it the machine came to life, noisily printing out the start-up page. My son squealed in delight, and it was all I could do to keep him from touching the on/off switch for the rest of the conversation.

"Well that's an appropriate start!" Harry laughed, looking sideways at Joshua. In an attempt to understand how books like *The Purgatorio* might be widely distributed, I had called Harry to see if he knew about medieval printing practices. And I was not disappointed, for Harry understood a great deal not only about clocks but about printing too; indeed, it seemed that he was familiar with most of the technologies of the Middle Ages. "Most people think about Gutenberg when they think about the printing press," Harry explained, and, as my brow wrinkled, he said, "There was a form of printing available to writers in Dante's era. It was called block printing, and I have this record . . ." here Harry pulled out one of those dusty old books ". . . showing that it was practised in Ravenna as early as the thirteenth century. And China and India were doing a form of it several hundred years before Christ!"

Harry explained that block printing, often called relief printing, was simply a block of something solid whose surface is partially cut away, leaving in relief the desired printing design. All one had to do was to apply ink to the block, manipulating it in such a fashion that only those raised areas got the pigment. Then, you simply pressed the block down onto the surface to be printed. "It's a far cry from *that* little device," Harry pointed at his photocopier as Joshua made another lunge for the on/off button. "Although this old technique is still in use, these days we call it letterpress printing." Harry quickly finished as Joshua started fidgeting against my restraining hand. I thanked Harry, cut short our meeting just as the professor was pulling out another book, and left in a hurry.

This explanation of medieval printing made a great deal more sense to me than Harry's discussion of his medieval clocks. But I bring up the notion of duplicating copies and on/off switches for the same illustrating reasons we talked about medieval clocks. To understand the genes behind the biological clocks of circadian rhythms, it is important to explain a specific *genetic* copying mechanism, one reminiscent both of block printing and modern photocopiers. The copying also involves on/off switches in genes. When the genes are toggled, they actually make copies of themselves, and from this copying we get our circadian rhythms. To see how medieval and modern printing practices illustrate a genetic mechanism, we will first need to discuss two pieces of molecular information. Then we will return to our discussion of circadian rhythms and observe how gene activity fuels the process.

As we explained in our chapter on lust, awakening a gene really means getting it ready to make messenger RNA, also known as mRNA. The purpose of most genes is to encode the information necessary to make proteins. Genes can't get out of the nucleus, and the protein-manufacturing sites cannot get into the nucleus. This means that some mobile form of communication between these two cellular areas is needed in order for protein manufacture to occur. As we learned, the cell has indeed made such a mobile entity, called mRNA.

A gene becomes activated when it is "copied" into mRNA. As in block printing, the gene serves as the template for the mRNA's construction. A series of enzymes which we will discuss shortly literally "feel" the imprint of the gene, like a piece of wood on paper, and make a copy of what they see. The mRNA leaves the nucleus, finds the manufacturing site, and a protein is soon made.

To understand a little bit more about circadian rhythms, we have to add to our knowledge of the genetic code, as well as gene activation. There is a puzzle that requires some figuring out regarding the constituent parts of both proteins and genes, and to see what I mean consider the following:

As discussed in Chapter Two, proteins are made of individual subunits called amino acids. Don't let that last word confuse you. These acids are not the burning kind of chemical found in a standard car battery, and the cell survives quite nicely with these acids tucked inside them. That's fortunate, because there are a lot of amino acids within a given cell. A typical protein can have hundreds of amino acids, joined together like boxcars on a train. Or perhaps I should say "train wreck" (in truth, the average protein has this line of amino acids twisted all around itself, making the protein look like some superhero had taken a long freight train and turned it into a pretzel).

The presence of amino acids gives us one part of the puzzle. To understand the rest of it, we need to look at the constituent parts of genes. A gene is also made of tiny subunit molecules, which we term nucleotides (we also use the word "bases" when we speak of nucleotides). Like amino acids, these nucleotides – these bases – are joined to each other in a linear arrangement. There are four kinds of nucleotides, which scientists label with the letters A, T, G, and C. A gene can easily have thousands of nucleotides, strung out in a line like beads on a string. A typical gene sequence might be AGGGACTAGGCCCAGAGCTA and so on, for a thousand or more bases.

The nucleotides give us the other part of the puzzle. When a gene becomes "active", it block-prints out a mRNA copy. This copy is also constructed of nucleotides, and in a linear arrangement reflective of the gene that made it. And therein lies the full puzzle. A mRNA is supposed to be able to talk to the protein-manufacturing site and instruct it to make a protein. But the protein itself is not made of nucleotides, but rather consists of amino acids. How does information embedded in nucleotides end up becoming the instructions for something made of amino acids?

Finding the answer to this question is one of the greatest achievements in all of biology. Obviously, some kind of translation protocol must be employed to read the nucleotides, and, once read, to specify the amino acids required. This protocol is termed the genetic code – in fact the word "code" is used to specify that some kind of language translating is indeed required. As you might expect, there is not only a way for nucleotides to specify the "words" that make up amino acids, there are also nucleotides that give us punctuation (it is a code, after all, in need of many of the requirements of normal language). Here is how the protocol works:

Every three nucleotides in a mRNA specifies an amino acid . . .
. . . or some kind of punctuation.

When a mRNA reaches the protein-manufacturing site, mechanisms within the site read the nucleotides in the mRNA in chunks of three. Most of the time the chunks will specify a given amino acid. For example, there is an amino acid named threonine. It is specified by the nucleotide trio ACG. The protein-manufacturing apparatus will read the ACG, and then go get a threonine. Let's say there is another trio of nucleotides, say CCC, laying right next to ACG. The CCC trio specifies not threonine but another amino acid, one called proline. The apparatus will read CCC, go get a proline, and join the proline to the threonine.

This joining can easily happen several hundred times before one protein is made. The manufacturing apparatus starts at one end of the mRNA, looks at the first group of three and then starts building. When the first group of three has been read, the apparatus ratchets past them and reads the next group of three, finds the appropriate amino acid, hooks it up to the previous amino acid, and then ratchets over again. Eventually, an entire protein is made. We give these chunks a special name: every trio of nucleotides is called a "codon", an apt word describing the coding function.

Note that in the previous paragraph I said, "most of the time the chunks will specify a given amino acid". Sometimes the chunk-of-three does not specify an amino acid, but instead specifies punctuation marks. What do I mean by punctuation? The concept is fairly simple, really, and has a metaphor in language. Consider the following English sentence:

Harry sleeps most of the afternoon.

You can tell when the sentence you are currently reading either stops or starts by simply noting two things: the capital "H", which begins the sentence, and the period at the end of the sentence. These punctuation marks give you a bracket around the information being conveyed, hopefully allaying confusion as to its message.

A given mRNA has the very same start and stop messages embedded in its nucleotides, and for the same reason. There is what is called a start codon, for example, telling the protein-manufacturing apparatus which trio of nucleotides to start reading. There is also a stop codon, telling the protein-manufacturing apparatus when to stop. They work just like capital letters and periods. The information inside a given mRNA is thus bracketed between start and stop codons, also hopefully allaying any confusion as to its message.

One of the most interesting aspects of mRNA is its ability to survive once it has been made. We discussed in the lust chapter that the mRNA must leave the nucleus and enter the world of the cytoplasm in order to make a protein. It turns out that the cytoplasm is a very hostile place for most mRNAs. There are molecules lurking behind every cytosolic corner whose sole function in life is to destroy specific mRNAs. It is something of a war, and the ability of an mRNA to survive is a very important factor in determining whether a protein is made or not. If the mRNA can get past the lurking destroyers, it will find the protein-manufacturing site and become translated. But if the mRNA doesn't survive, no protein will be made. Believe it or not, many cells use this mRNA stability as a way to regulate the amount of protein made in a given cell. If the cell employs a lot of destroyers, the gene may make a tremendous amount of mRNA, but only a tiny amount of protein.

The second piece of information

With this idea of the genetic code in mind, we are ready to talk a little more intelligently about circadian rhythms in fungi and in human beings. As we mentioned at the beginning of this section, we need to review briefly a second biological concept before we do so. This is not translating a given mRNA, but waking up genes to make the RNA in the first place.

As you know from reading Chapter One, most of the genes inside a human being are asleep. Awakening a given gene means allowing it to become block-printed into mRNA so that a protein can be manufactured. A typical human cell has over one hundred thousand genes capable of being awakened, and so the question becomes one of regulation: who decides what gene gets turned on, and when? That's an important question. Recall that each cell in a human being contains genes for *every* tissue in the human body.

The answer to the question about activation comes from understanding the nature of DNA sequences that often lie just upstream of the gene. It turns out that every gene in every cell contains an on/off switch, just like a modern photocopier contains an on/off button. This switch is called a promoter, the word taken from the idea that this important upstream DNA "promotes" activation. But don't be fooled by that word, it is truly an on-and-off arrangement. Detecting the presence of a promoter gave researchers an idea as to how genes could be activated. Indeed, most genes discovered to date have promoter regions.

How does the gene use the promoter to become active? Is there some kind of molecular finger, like Joshua's hand, that presses the switch and turns on the gene? And is there a copy machine that will respond once the button is pressed?

The answer to both questions is "yes". There do appear to be proteins that can actually land on this switch and turn it to the on or to the off position. We call these proteins transcription factors. Once the switch is "thrown", the gene becomes active, something like turning on a photo-copying machine. The gene then gets copied into mRNA as a result. There is also a copying mechanism that is functionally separate from the transcription factors. The copying apparatus is also made of proteins. We call these groups by the collective name of RNA polymerase. When transcription factors bind to a promoter, they will coax the photocopying RNA polymerase to come and make a copy of the gene into mRNA. In summary, the two steps are: (1) binding of transcription factors to a given gene (this selects the "target" sequence that is to become activated), and (2) binding of the copying mechanism to the target, so that a mRNA can be produced.

Is the complete job description of a transcription factor simply to turn genes on? The promoter, after all, has an "off" function as well. How is that off function used? It turns out that a transcription factor can turn on a gene that is sleeping, as we just discussed. But many transcription factors can bind to a gene that is already active *and then turn it off*. In other words, a transcription factor can repress activity as well as stimulate it, which means there are true on/off capabilities. When transcription factor proteins perform this inhibiting function, they are logically termed repressors. As you can see, transcription factors turn out to be versatile little devices. While the actual situation inside living cells is a bit more complex than I intimate here, the notion of activation and repression is both accurate and resident within the nuclei of all creatures.

Regardless of whether one talks of on or off, there is an interesting question that lies at the heart of the regulation of transcription factors; indeed it is actually very relevant to our discussion of circadian rhythms. Transcription factors are themselves proteins, as mentioned above. That means they are also encoded by genes, and since every gene has an on/off switch (a promoter), there must be transcription factors that can turn *them* on or off as well. Does that sound confusing? The question is: who regulates the regulators? And then, who regulates the regulators who regulate the regulators? You can keep going on this question *ad nauseam*.

Answers to these convoluted interrogations can be found by examining genes that govern circadian rhythms in cells. The notion that layers of regulation exist in circadian neurons lies at the heart of the mechanisms that guide the independent pulsations of these interesting cells. Interestingly, we did not gain our understanding of the regulators by first studying human neurons. Indeed, the insights came from an unexpected source, a bread mold, of all things, which goes by the complex name of *Neurospora crassa*. This mold is a creature belonging to a class of organisms whose biology used to have very different explanations. To understand what I mean by that last sentence, I would like to describe one of the most interesting papers I have ever read in the published scientific literature. Its subject did not deal with *Neurospora* however, but rather with fluid-filled vials and religions. And miracles. How this all fits together is described below.

The stigmata of iron

"You've gotta read this, John!" A fellow scientist and friend roared to me from one side of the lab. He intoned, "This is gonna make your day". My friend was busy reading a scientific journal, and talking to me about a most interesting article. Its subject was religious miracles of all things. His enthusiasm for my input came from the fact that I am a religious person (he is not), and his laughter was reminiscent of a thousand philosophical conversations in which we have engaged and continue to have to this day. He literally threw the magazine across the room.

"Wow," I responded as I read the title. And I am sure its subject made my friend's day. The letter was talking about the scientific explanation of what was originally thought to be a miracle. It seemed that there was this fluid-filled vial in a cathedral, and at certain times of the year the fluid in the vial appeared to turn red. The faithful had long believed this transformation to be reminiscent of Christ's blood. The article was offering a more rational explanation, something invoking seasonal environmental conditions and precipitating iron oxides. While not dismissing miracles outright, the powerful head of science once again appeared to be smashing the few superstitions left in the twentieth century. It was grist for more discussions between me and my colleague. I threw the magazine across the lab back to him.

In medieval times, there would of course be no explanation such as iron oxide precipitation to describe this miracle. Things like stigmata (the bleeding of objects or people that looked like the wounds of Christ), weeping statues, and preposterous medical properties of relics, all these superstitions were considered to be normal expressions of Divine Presence within the natural order. Even naturally occurring phenomena did not escape superstitious explanations. Consider that bread mold *Neurospora crassa* I mentioned previously. Much like the iron-oxide "miracle" in the cathedral, *Neurospora* can also periodically turn crimson (in fact, at a certain stage in its life, it actually cycles its color, turning red and then white every twenty-four hours). In the Middle Ages, such natural occurrences would not be described in physical terms, but, like the fluid-filled vial, in spiritual terms. Cycles of red and white were the miracles of God, sent by Him to remind us of certain spiritual principles, as if the creatures were having their own circadian stigmata (a white flower might stand for the purity of The Virgin, for example, the red for the blood of the crucifix and so on).

These days we know that bread molds like *Neurospora* do not turn red on a regular basis to remind us of the blood of the cross; indeed, the lesson one learns from the vial is that physical phenomena have physical explanations. But having ruled out superstition, what might be the physical reason for red-and-white bread molds – the remarkable twenty-four-hour periodicity of *Neurospora*? Scientists knew they had in this fungus a powerful experimental model with which to study circadian rhythms. And so they began an investigation which was eventually to shed light not just on lower fungi, but on every creature that possesses a circadian cycle. With the concepts of the genetic code and transcription factors in tow, let us return to the world of biorhythms, focusing first on this bread mold. We will see how these mechanisms shed light on the way creatures like ourselves respond to our days and nights.

A world of genes in a fungus

It all has to do with the choices they make about their sex lives, literally. Unlike humans, the bread mold *Neurospora* has the choice of several ways to reproduce. It has the familiar sex cycle, complete with the interaction of opposite mating types, jumbling of genetic information and the creation of diverse children. But *Neurospora* can also reproduce without sex. In a process termed conidiation, the fungus can make a series of hardy spores, which can then germinate to form the ruby gossamer-like structures we find on old bread. It is this germination that gives the fungus its red color.

What interested circadian scientists most about *Neurospora* was not its eating habits, however. As mentioned, this red color appears regularly in a circadian cycle. If one allows the fungus to grow at one end of a test tube, it will shoot out little feeler-like structures, travelling along the inside of the glass, producing red-like spore structures every twenty-four hours or so. In other words, the process of conidiation – the making of spores – exhibits circadian characteristics. This is important for our discussion because *Neurosporas* have no complicating brain tissue. Yet the fungal life-cycle is fully responsive to day/night cues. Scientists were very excited because *Neurospora* is a very simple organism, well characterized, and one that lends itself easily to genetic studies. Of course they decided to investigate.

After a great deal of work in many laboratories, the genetic basis of this fungal rhythmicity is becoming clear (see Figure 16). *Neurospora* has a very interesting gene called *frequency*, or *frq* gene, for short, pronounced either "frik" or "freak", depending upon who is giving the seminar. (As an aside, please note that *frq* is italicized. The name of the gene is always italicized, while its protein is not. This helps distinguish, in short-hand fashion, precisely which biomolecule one is talking about).

The *frq* gene product is a transcription factor with some very interesting properties. It is one of the very first genes ever discovered whose job

Lessons from a bread mold

We have learned a lot about circadian rhythms from the common bread mold Neurospora crassa. *One of its best studied molecules is called* Frq, *explained below.*

frq REGULATION

The promoter region of the *frq* gene is tightly controlled, providing the region where the mold' biorhythms are established. Regulation of the *frq* gene can be divided into four steps.

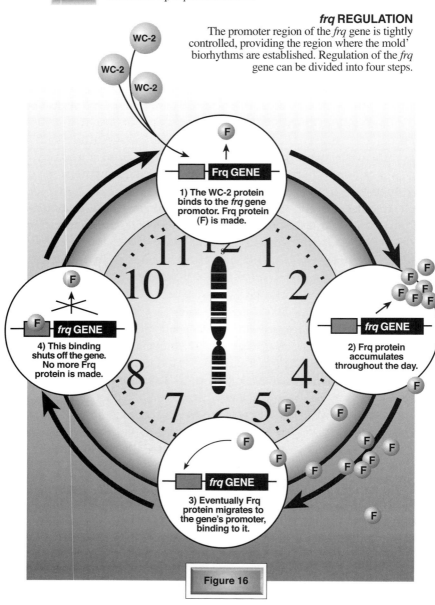

1) The WC-2 protein binds to the *frq* gene promotor. Frq protein (F) is made.

2) Frq protein accumulates throughout the day.

3) Eventually Frq protein migrates to the gene's promoter, binding to it.

4) This binding shuts off the gene. No more Frq protein is made.

Figure 16

description fits the bill of a "rhythm oscillator". By its cues, the fungus "knows" to start making reproductive structures at a certain time of the day. Frq proteins work in a fashion similar to that quartz crystal in the fancy Swiss watch we discussed earlier.

How do we know that the *frq* gene and its product Frq work as a governing oscillator? Two pieces of data demonstrated its time-piece activity. The first had to do with *frq* gene activation; the *frq* mRNA levels follow a circadian rhythm, peaking in the early morning hours. Secondly, these levels respond rapidly to the presence or absence of light, demonstrating an environmental detector function. (*Neurospora* possesses pathways sensitive to specific wavelengths of light, one of the best characterized is a pathway that responds to blue light, a pathway with which the *frq* gene is deeply involved.) The *frq* gene was thus found to perform two hallmarks of circadian biology: it works as a clock that both measures and displays time.

The biggest surprise concerning the *frq* gene is the characteristic that has ended up making the most sense from a mechanistic point of view. Frq protein is indeed a transcription factor. Everybody wondered what its targets would be, and the answer to that question gave researchers the aforementioned surprise. On what gene does the Frq protein work? The Frq protein actually binds to the *frq* gene promoter– *and turns it off*. You did not read that wrong. When there is enough Frq protein made, the protein marches back into the nucleus, finds the very promoter that made it, binds it and turns it off. In other words, the *frq* gene product is a repressor of its own activity.

If you think about it for a second, the whole thing acts something like the kind of negative feedback system we described in the chapter on gluttony. The *frq* gene activity peaks in the morning, and, when enough Frq protein is made, *frq* gene activity begins to decline. Eventually, it is turned off, and one aspect of the clock is explained.

Note that I said only one aspect is explained. The Frq negative feedback loop helps us understand how the gene can be turned off. But what turns the *frq* gene on in the first place? Understanding both positive and negative aspects of regulation is important in closing the loop, and giving a complete explanation of the clock.

The answer to this question has recently been explained. Two more genes have been isolated, called *wc-1* and *wc-2*, that seem to serve as alarm clocks for a sleeping *frq* gene (the letters wc stand for "white collar", a description of what *Neurospora* looks like if the gene has been mutated). One gene product in particular, the protein encoded by the *wc-2* gene, appears to be a stimulator of *frq* transcription. When the WC-2 protein lands on the promoter of a sleeping *frq* gene, the *frq* gene wakes up and starts making its own protein.

With the isolation of these two genes, we perhaps have the beginning explanations of a closed-loop circadian system, the first ever uncovered. WC-2 wakes up the *frq* gene, and the subsequent Frq proteins go to work providing timing functions for the mold. Eventually so much Frq protein is made that some of it begins to wander around the cell. Some of the protein actually travels back to the *frq* gene, finds the promoter and turns it off. Frq protein levels then start declining, and continue to do so all day. Next morning, WC-2 proteins come along and turn the *frq* gene back on. Frq protein levels rise once again, and, voila, we have a circadian cycle.

Since WC-2 is a protein, obviously encoded by a gene, the next question becomes: what turns on the *wc-2* gene? As we have discussed, this kind of regulatory question can be asked *ad nauseam*; it is a kind of chicken and egg which-gene-gets-turned-on-first type question. As of this writing, the cutting edge is currently at the level of the *wc-2* promoter (see Figure 16).

At this point, you may be thinking to yourself: its okay to know about this Frq stuff if you like bread mold, but what about *us*? What does studying *Neurospora* have to do with understanding how the sleep cycle is regulated in human beings? The answer is plenty. As we have seen in previous chapters, researchers have often had to use animal models in order to find the starting place for research in human beings. It turns out that *Neurospora* gave us some great clues as to where to begin looking in creatures much more complicated than bread molds. Before we start talking about human beings, however, we have to look at one more organism, the one that took direct hints from the bread mold and really gave researchers the clues they needed for what goes on in humans. Perhaps humiliating for us, that creature happens to be the fruit fly. In our quest to understand the genes in human beings, it is to this humble insect that we turn next.

What's going on in insects?

The common fruit fly has been a research staple for many geneticists literally for decades. It is completely amenable to genetic manipulation, and as genetic engineering techniques have come into regular practice, researchers have found that a surprising number of genes have analogous functions in mammals. The insect has a distinct advantage over bread molds like *Neurospora*, in that fruit flies have neurons. And it turns out that these insect neurons possess circadian rhythms just like ours, hence our reason for bringing it up here.

There are many gene sequences involved in the regulation of circadian rhythms in fruit flies. The two genes I would like to discuss in our brief survey are called *per* (short for period) and *tim* (short for timing) (see Figure 17). The role that these genes play as pacemakers in the life of the fly are straightforward: in the adult, the *per* gene is deeply involved in its activity during the day (so-called locomotor activity), and also certain developmental aspects of its early life. The *tim* gene is chiefly involved in the regulation of the *per* gene. As you can see, the cooperative aspects of these genes and their products help produce the circadian rhythms seen in this creature. It is our task to find out how they work together, and then try to relate this to both bread molds and people.

As with all genes, both *per* and *tim* possess promoter regions. The genes become transcriptionally active through interactions of certain molecules with these promoters (more on that in a minute), and the mRNAs pass from the nucleus and into the potentially hostile cytoplasmic environment. If they survive, both mRNAs will be translated into proteins. Now present in the cytosol, the level of cooperation begins with these proteins.

It turns out that the Per protein possesses a specifically sticky region on it called the PAS region (it's actually a group of amino acids, like someone had inserted fly paper into the protein). This sticky region allows the Per protein to bind to certain molecules, if the right conditions are present. And to what molecule does Per bind? It forms an association with our other protein, the Tim protein, creating an interesting duet of a molecule scientists like to call heterodimers. How do the proteins "know" to find each other and stick? The answer involves the presence of light. If light is encountered, the proteins get the energy they need to form their fancy heterodimer. And then, in a process that also appears to be mediated by light, a number of copies of the duet leave the cytosol and re-enter the nucleus.

Once inside the nucleus, the Per/Tim protein complexes do something very reminiscent of the Frq protein in bread molds: they look for promoter regions. The heterodimer finds the *per* gene promoter, binds to it and shuts the *per* gene down (sounds familiar? the Frq protein does a similar thing to its gene). Another copy of the heterodimer finds the *tim* promoter, binds to it and shuts down the *tim* gene. Soon there are no more mRNAs of either gene being made, and the supply of new protein decreases steadily. For the second time in our discussion of circadian rhythms, we have encountered a powerful negative feedback loop. This inhibition is mediated somehow by the presence of light. Without it, the proteins would never

What we've learned from fruit fli

The gene products Per *and* Tim *control circadian rhythms in many*
animals. Best understood in fruit flies, the interaction of these two
proteins provides a light-sensitive molecular clock for the creature
harboring them. Here's how they work.

A COOPERATIVE ADVENTURE

Both Per and Tim proteins work together to provide circadian information. The Per protei
contains a region called PAS, a sticky area capable of binding to the Tim protein (though onl
if light is available). Once bound, this heterodimer can re-enter the nucleus, bind to both *pe*
and *tim* gene promoter regions, and shut them down. This feedback pathway is shown in fiv
steps below

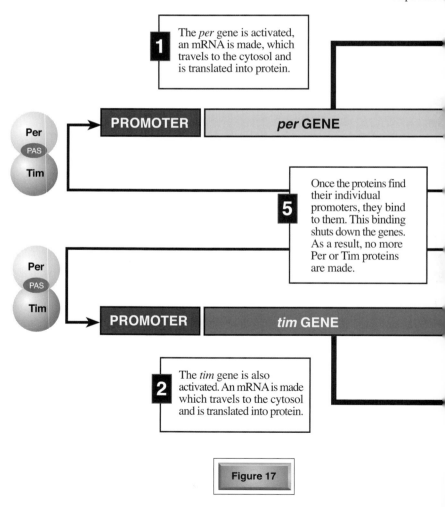

1 The *per* gene is activated, an mRNA is made, which travels to the cytosol and is translated into protein.

Per
PAS
Tim

PROMOTER   *per* GENE

5 Once the proteins find their individual promoters, they bind to them. This binding shuts down the genes. As a result, no more Per or Tim proteins are made.

Per
PAS
Tim

PROMOTER   *tim* GENE

2 The *tim* gene is also activated. An mRNA is made which travels to the cytosol and is translated into protein.

Figure 17

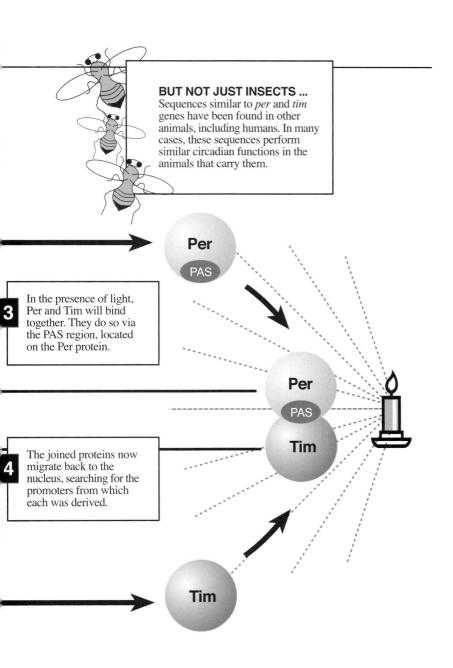

BUT NOT JUST INSECTS ...
Sequences similar to *per* and *tim* genes have been found in other animals, including humans. In many cases, these sequences perform similar circadian functions in the animals that carry them.

Per
PAS

3 In the presence of light, Per and Tim will bind together. They do so via the PAS region, located on the Per protein.

Per
PAS
Tim

4 The joined proteins now migrate back to the nucleus, searching for the promoters from which each was derived.

Tim

enter the nucleus; indeed, without the light, they would not even stick together. Sensitive to the presence of light? That's what circadian rhythms are. This system is very sensitive to whether it is day or night outside, and makes the creature respond to the Sun by turning genes on and off. It is a true circadian rhythm system, now fleshed out in terms of molecules and genes.

There is another level of regulation worth mentioning in our discussion of these fruit fly gene products. You recall that the cytosol can be a fairly hazardous place for many mRNAs. Once they leave the nucleus, the mRNAs are often destroyed before they can reach the protein-manufacturing site and become translated. There appears to be a regulatory mechanism for destroying the *per* mRNA. It has been found that this particular mRNA has a very short life if left alone in the cytosol of fruit fly neurons. The fact that this transcript is so unstable – especially when compared to other transcripts that do survive well – means that the *per* mRNA is *specifically* targeted. It certainly would be a way to control the amounts of Per protein available for binding to the Tim protein. This mechanism might represent one more gear in the ticking mechanism of the fruit fly's circadian rhythm.

Per genes and *tim* genes make a nice circadian story, showing that the negative feedback loops observed in bread molds really do occur in insects. And while it is convenient that *Neurospora* provides an experimental framework for fruit flies, we still don't know very much about humans. Can the mechanisms found in Per and Tim (and indirectly Frq) give researchers hints as to what occurs in more complex animals, like us mammals? The answer to that question turns out happily to be yes. But the researchers, still bound by the necessary ethical constraints on human experimentation, had to use the mouse to further the story. Always risky of course, but there's a principle in looking at genes I would like to relate before we discuss what was found in mice and humans, a principle that makes looking at animals worthwhile. The principle is illustrated in my third and final trip to see Harry, which consisted only of returning a book I had borrowed.

The gargoyles of Rouen

After Harry had told me about the fourteenth century clock in the French city of Rouen, he gave me a book on French medieval architecture and told me to leaf through its pages. I went back to his office to return it.

"Did you like it?" Harry beamed, peering over his horn-rimmed glasses, struggling to light a cigarette.

"I did indeed," I said, waving away at the smoke. "I thought the gargoyles in Rouen's cathedral were about the most interesting things I have ever seen."

"Kind of a botanical garden, menagerie and fantasyland all at once, eh?" And Harry wasn't lying. "Before Rouen, the twelfth century cathedral reliefs were really boring," he continued. "They were interested only in the structural lines, not in any sculptural artistic ideas. The thirteenth and fourteenth centuries, that's where you get the exuberance, that's where you get the delight." He absentmindedly began leafing through the book. "This is my favorite page."

He stopped at the place where I too had felt the greatest whimsy, the Gothic quatrefoils in Rouen cathedral. There's a figure that's obviously a doctor, studying a vial of urine, but the sculptor has made the doctor half man, half goose. There's a figure of a contemplative philosopher with the head of a hog. There's even a music professor giving an organ lesson to a Greek centaur, and the professor is half human and half rooster. "That reminds me of someone in our department," Harry chuckled almost to himself, and blew a smoke ring. I laughed too, and as I left Harry's office it dawned on me why I felt the reliefs to be so memorable. The whimsy came because of the skill of the artist. The sculptures looked very authentic, very *real*, down to the muscles and the feathers and the hooves and the fingers. Because I could immediately identify the structures as belonging to one creature or the other, I was able to make inferences about their functions; had I known more about medieval bestiaries, I might even have been appreciative of a little medieval satire.

It is this principle of the recognition of the familiar, and of inference of function that directly relates to our story about circadian rhythms and genes. In a moment, we are going to examine some genes in mice that look very much like genes in fruit flies, almost as if the nucleus comprises a molecular gothic statue that's half rodent and half fruit fly. What I mean by "look like" has to do with similarities in the genes, their nucleotide sequences (those A's, G's, C's, and T's we discussed earlier) and the amino acid sequences of the proteins they encode. Whenever we scientists encounter similarities in nucleotide and/or amino acid sequences between two genes, we get very excited. That's because two genes that look similar often have similar functions. We call such similarities conservation of function, a concept upon which we promised to elaborate way back in our gluttony chapter. Here's an example.

Let's say I was hunting in a medieval forest and came across a mammal I had never seen before. Let's say that the mammal was a normal non-whimsical creature, had a face whose central portion was occupied by this moist black protuberance with which air was taken in and expelled. Based on experiences with other mammals, I might reasonably conclude that the creature had a nose, and probably used it both for respiration and for smelling things. Why did I infer these facts? Because the structure looked like other noses that I had seen before on known mammals. Thus, even though I had never encountered this creature, I still could make some fairly educated guesses about some of its anatomical functions based on structural similarities.

This recognition of similarity is the reason why the gothic relief sculptures were so whimsical and humorous, for I could recognize both animal and human in a single form. And it is this inference of function based on similarity of structure that is the root reason why scientists get so excited when they find conserved genes, especially if these genes exist in creatures who, from an evolutionary point of view, are only distantly related. In the next section, we are going to talk about mammals, and we will discuss a gene found in mice that looks very similar to genes found in fruit flies (we will also find that the gene appears to exist in humans as well). Thus, even though we cannot do experiments on humans, we can begin to ask better research questions based on what we find in mice, especially if we find and isolate sequences that are common to many different types of animal because they have been conserved in evolution. It is to these extraordinary genes that we now turn.

And onto mammals

For the reasons stated above and for other motives, it is great to work on circadian biology with mice. They have an SCN after all – that area of the brain, as we have discussed, which provides most of the pacemaking functions of the animal. It is an exquisitely precise timepiece. If the rodents are kept in total darkness, they will still keep an exact 23.7-hour circadian cycle consisting of running and resting, running and resting, similar to us humans. However, there are mutations in some mice that disrupt this exact little time-piece. It was found that if a mouse had one copy of the mutated gene, their periodicity was expanded from 23.7 hours to 25 hours. If a mouse got two copies of the mutant gene, the periodicity began to break down completely. Initially, these two-gene carriers expanded their cycles to 28 hours, but after two weeks in total darkness the mice lost their rhythmic patterns completely. In other words, the mutation shut down their cycle.

Of course the scientists immediately became obsessed with isolating this powerful circadian gene, especially because it was not from a fungus or an insect, but from a mammal. After a considerable effort, the gene was isolated and called, appropriately enough, the *clock* gene. This was quite an achievement, for it was the first circadian gene ever discovered in a mammal. Eventually a few other circadian genes were discovered in mice, including two that looked very similar to certain regions of the *per* gene of the fruit fly. Indeed, this similarity is demarcated in their names, *mper1* and *mper2*, the "m" because they come from mice.

The next set of questions focused upon their interactions; as you might expect, their similarities to the other genes gave scientists a hint as to what kinds of questions to ask. It was found that both *per* genes are expressed in the nerve cells of the SCN (the SCN, that region of the brain containing the pacemaker neurons in humans, performs a similar function in mice). Both genes show a circadian pattern of expression when the mice are immersed in complete darkness. Curiously, however, they are not turned on at the same time in the lightless environment, but are instead about four hours out of phase.

It was discovered that the gene expression differed greatly when the researchers turned on the lights, and it is this difference that gave them hints as to function. The *mper1* gene is rapidly turned on in the mouse's brain in response to daylight, but the *mper2* gene is completely unresponsive. What does that mean? No one knows. Some researchers believe that *mper1* is the actual pacemaker gene, or at least the component in a complex jumble of molecules whose job description is to aid in the overall response to the presence of light.

Whereas the *mper* genes give functional hints based on their similarities to insect molecules, the *clock* gene is a bit harder to pin down. Researchers knew about its central role in the sleep cycle because of the crippling effects on circadian behavior when the genes were mutated. You can even go in the opposite direction and put more copies of the *clock* gene into the mouse than it is supposed to have. When that happens, the circadian rhythm in the over-stuffed mouse "speeds up". This gives strong evidence that the *clock* gene really is the central oscillator; the more the system gets built up, the faster the mouse zips around the clock.

How then do the *mper* gene products and the *clock* gene products interact with one another? As of this writing, the answer to the question remains a mystery. But the structures give researchers ideas, fortunately, with clues conveniently provided once again by the bread molds and the insects. Remember those sticky PAS regions found on the Per protein in the fruit fly? Those were the areas that allowed the Per protein to bind to the Tim protein, making that regulatory duet so important to the insect's circadian rhythm. It turns out that *clock* genes in mice have a region that looks just like that PAS sequence. When the researchers looked at the *mper1* and *mper2* genes in the mice, they discovered PAS sequences as well. Does this mean that Per and Clock proteins interact with one another? No one knows presently. But the fact that they interact in fruit flies gives the researchers hints as to which experiments should first be performed.

It also gives researchers direct hints as to where to look in humans, and important strides have already been achieved. Recently, for example, a gene has been found on human chromosome 17 that looks a great deal like the *per* gene of fruit flies and mice. The gene has been dubbed *rigui*. This *rigui* even has a flypaper PAS domain, which makes the researchers think a lot about *tim* genes, molecular duets and conserved mechanisms. Perhaps not surprisingly, the gene has also been discovered in the mouse, where its expression has been found to rise and fall in the familiar twenty-three- to twenty-four-hour circadian fashion.

Summarizing these molecules

Taken together, research into bread molds, insects and us has yielded great insights – and surprising unity – in our attempts to understand how individual neurons react to the Earth's days and nights. It is the most intimate way we can look at the internal clocks inside cells, and, by inference, into the creatures that carry them. The next task is going to be the hardest, for the interior clocks are only part of the picture. At present, we don't really know how the gene-supervised pulsations in many thousands of cells work collectively to create coherent circadian information vital to the rest of the body. We know that cells march to "different drummers" regarding their own firing patterns, but how the specific signals are collectively understood at this point remains a mystery.

The incompleteness is hampered for another reason. Isolating *frq*, *per*, *tim*, and *rigui* genes represents enormous strides, but there are many other genes involved not mentioned on these pages that serve both to inform and complicate the overall picture. Most frustrating of all, there are strong hints from the research laboratories that we have only isolated a few of the most important genes involved in circadian biology. Many – some researchers would say *most* – await discovery, and their roles must be well characterized before we even attempt to create an integrated biological picture of biorhythms. This is especially true when we talk of human beings. Once again, we come to the same hard lesson we encountered in the previous chapters. We know enough to create kilograms of testable genetic experiments, but we don't know enough to state definitively how genes work.

Taken together

From SCN to melatonin, from *frq* to *per*, it is obvious that a great number of strides have taken place regarding the biology of circadian rhythms. The unity of the mechanism between even the most disparate of creatures (*per* equivalents are found in both humans and flies!) is both amazing and somewhat predictable. Terrestrial creatures all live on the same planetary surface, and so all are exposed to the twenty-four-hour cadence. Where it is going to be in space and time – and what level of activity will occur in a given timeframe – is one of the most important daily survival decisions a creature faces. The organism only has two stages upon which to play out the biology, either in the light or in the dark. The neurons that control a creature's reactions to these stages are important indeed.

We have not always thought of our alternating cycles of drowsiness and activity in terms of biological phenomena, of course. One of the reasons why we are briefly reviewing the history of mind/brain debates, as well as addressing the concept of human consciousness, is to show how revolutionary a lot of this neurobiological thinking is. In the last part of this chapter, I would like to use circadian rhythms to return to our discussion of the history of minds and brains, as well as introduce yet another ingredient in the definition of human consciousness. We shall end the chapter by returning to the idea that while the circadian rhythm is not an emotion, its disruption can generate a wide variety of moods. This reveals a lot about the nature of pacemakers, and uncovers something very important about the scientific study of behavioral/emotional phenomena.

Back to minds and brains

Let us continue our brief review of the mind/brain debate by quoting a verse taken from the Fourth Cornice.

> . . . *love, which is a spiritual motion,*
> *Fills the trapped soul, and it can never rest*
> *Short of the thing that fills it with devotion*

This verse is right in the middle of the speech Virgil is giving to Dante when they first arrive at the Cornice. Its topic concerns love, free will, and boredom (Dante falls asleep during the lecture). As is true of most of these discourses, a great deal is revealed concerning medieval attitudes to and perspectives of souls and bodies, and, by inference, minds and brains.

Of course, medieval ideas about human cognition and emotions did not wrestle with the notions of mind and brain, as is clearly revealed in this verse. As we have seen, the perspective of the Middle Ages was taken somewhat from Plato, who held that a corporeal soul – a "trapped soul", to use Dante's words above – inhabited an earthly body. (You might recall that church thinking in the Middle Ages could also take ideas from another Greek. Democritus, who held that *everything* was material in nature, may have been rejected because his world-view was not as easily reconciled Biblically). As mentioned on p. 101, it wasn't until Descartes came along, with his famous dualism, that the debate began to take its more modern shape. The problem had to do with the interface between our bodies and our souls; specifically, how they reacted and informed each other of life's events. Descartes' odd answer, positing the role of the pineal gland as the place where soul met body or, in modern terms, where mind met brain, turned out to be unsatisfactory.

We learned in the last chapter that other philosophers took stabs at addressing the nature of the interaction, and that their solutions turned out to be equally fantastic. Leibniz said that the two never interacted at all, and that events which required both minds and brains to interact were separate coincidences, kept on track in the temporal world by God. Berkeley also avoided the question of interactions by saying that only the internal perceptions of mind drove reality. In his view, the essence of an object existed only if someone perceived it, and God held everything together because He, needing no sleep, continually perceived all of creation.

As you might expect, many people reacted in a skeptical way to these imaginative solutions of mind and brain interactions. By the time we get to the seventeenth and eighteenth centuries, a school called Materialism was starting to form. And its solution was another form of avoidance. Materialists such as Thomas Hobbes and de la Mettrie held that the solution to the mind/brain problem was to simply exile the soul from the body. In this view, reminiscent of the Greek Democritus, the mind could be explained mechanically. The body was simply a very talented mechanical device. In the words of de la Mettrie's *L'homme Machine* (1748):

Since the faculties of the soul depend to such a degree on the proper organization of the brain and the whole body ... the soul is clearly an enlightened machine.

As you might expect, this radical notion rocked the worlds of both religion and science. It was fueled in part by the great strides being made in engineering and science at the time. And de la Mettrie wasn't alone; Thomas Hobbe's points of view are illustrated by describing some of the technological achievements of the day. Consider the following quote:

Seeing life is but a motion of limbs, the beginning whereof is in some principal part within; why may we not say that all automata (engines that move themselves by springs and wheels as doth a watch) have an artificial life? What is the heart but a spring, and the nerves but so many wheels, giving motion to the body ...?

Materialists and the notion of consciousness

In many ways, the ideas of Hobbes (called by some the prophet of artificial intelligence) and de la Mettrie preface the modern ideas of neuroscience. These Materialists wrestled with the nature of awareness, and concluded that we were nothing but a fancy robot. The fact that many people nowadays find this idea repugnant shows that the debate about the nature of minds and brains is not yet over. But it pushed things in a testable direction. These days, the discussion is centering in on the hippocampus and neurotransmitters rather than on springs and wheels.

And that's where we are too, especially as we attempt to understand the biological nature of human behavior, especially emotion. As you know, we call such awareness these days consciousness, and we have been trying in this book to explain some of the modern views of its ingredients. This is important to our explanation of emotions because of our now-familiar model; we are positing that emotions occur when consciousness is aroused by some emotion system in the brain. Does our exploration of circadian rhythms inform the debate about mind/brain at all? Can *per* genes illuminate for us part of our model of emotions?

The answer to both questions is a tentative "yes", and the data are coming from the observation that mood can vary according to the length of time someone has been awake (for an intuitive example, just ask any parents of a newborn child!). Such an observation actually straddles the mind/brain question: do something to your body and you can do something to what Descartes would call your mind. The fact that these shifts concern awareness means that one of the ingredients of consciousness is deeply involved too. The data to which I am referring came from some of those sleep-disruption studies we discussed earlier, the kind where volunteer subjects are placed in isolation chambers and then various responses are measured. To see what I mean, let's consider an example. Here's what happened in one typical experiment.

A total of twenty-four people were involved in the study, living in isolation chambers for several weeks at a time. Their circadian days were artificially lengthened from the normal 24.2 hours to the experimental 28 or even 30 hours. In the words of the researchers, this was chronic "forced desynchronization" and allowed researchers to test human mood in the disruption of circadian versus sleep/wake patterns.

And what did the researchers find? The mood of a given person in the isolation chamber became positive or negative when they were awake according to the difference between their natural waking times and their actual waking times. Sound confusing? Suppose that the person in the isolation chamber was going to be awakened in what according to their *natural* circadian rhythm should be the late evening. That's a lousy time to be awakened. Their mood noticeably darkened, and it was always the worst between six and eighteen hours after waking up. Conversely, let's say that the person was awakened at what according to their natural circadian rhythm should be morning. These people naturally recorded higher mood levels throughout the day. Want to be in a good mood throughout the day? An important component is to keep synchrony between (1) the time your circadian rhythm tells you that you should be awake, and (2) the time you actually wake up. In other words, keeping certain neurons happy in one part of the brain (like the SCN) helps keep other areas of the brain from becoming moody. Thus, even though circadian rhythms do not necessarily constitute an emotion system *per se*, they can greatly affect the ones we possess.

Sounds a little bit like Democritus to me.

But what about consciousness?

From an intuitive point of view, the fact that having a disciplined sleep schedule is important for mental health is not exactly new information.

Being chronically tired can put almost anybody in a bad mood. What's interesting about these data is that they tease out the difference between issues involved in what time a person gets up compared to the inputs of their own natural clock. They even pinpoint critical hours for being ill-tempered.

Besides assessments of mood, other research has shown another obvious linkage: when you are sleepy, you are also not as aware of specific things (for example, details in a given task) as when you are awake. And while we are familiar with this association, scientists have long puzzled over exactly what a *level* of awareness is, and why amounts of sleep can control it. As you might expect, people who are interested in studying consciousness are also deeply interested in this question. And since consciousness is a large part of our own working model for emotions, we must briefly mention it here.

As you recall, we are attempting to characterize consciousness by describing some of its characteristics. In the last three chapters we have focused on memories, looking at long-term functions, short-term buffers, and finally working memory. In this chapter, we have been talking about sleeping and waking, both of which have their own inputs into our notions of consciousness.

There is a kind of awareness that we have not talked about yet, but still it has a lot to say about our knowledge of consciousness. Scientists like to call it an arousal system or, better, arousal systems. These systems, often called the reticular activating systems, are different to the memory processes we have been discussing. They have been localized to discrete regions in the brain. These systems even behave differently; they function like a neural drill sergeant, commanding our conscious attention to stay fixed towards some emotional stimulation when the need demands.

What do these arousal systems have to do with emotions? They are involved a great deal, actually. Emotional reactions usually come hand-in-hand with intense feelings of arousal. If we did not have these interesting arousal systems, we would still get the same emotional reactions, but they would then very quickly vanish. As you might expect, such emotional amnesia would not be a wise thing to experience. If our feelings of fear were only fleeting when we came face-to-face with a saber-toothed tiger, we might lose our overall motivation to run away. Many scientists believe that circadian rhythms, even though they do not constitute an emotion, can nonetheless deeply affect our arousal systems, and thus the emotions with which these systems interact. The data on the moods of the people in the isolation chambers suggest, among other things, this interesting linkage.

What is the link between these arousal systems and consciousness? At one time, it was thought that the brain areas involved in arousal systems controlled what was often called a "continuum of consciousness". This continuum started at one end with attention and alertness, went to relaxation and drowsiness, next to sleep and stupor, finally to coma. Nowadays the continuum idea has fallen out of favor, considering that even in sleep there are almost unbelievable bursts of neural activity in those brain areas involved in arousal. That does not mean that these arousal systems play no role in conscious awareness; indeed, everything that demands mental activity, such as memory, problem solving, and attention, are routed through these systems. They are thus a very important *ingredient* in our definition of consciousness. The regions that control these arousals just don't embody the entire notion.

Our brains

As we have seen in the above discussion, the biology of circadian rhythms informs the debate about our minds and brains. These day/night switches also affect the quality of one of the ingredients of consciousness. They also illustrate one aspect of behavior I would like to mention before this chapter closes.

In the last chapter, we used the concept of greed to talk about an important idea concerning how we look at emotions: a human category of psychological experience does not necessarily reflect biological reality. We mentioned that even though avarice is a feeling powerful enough to change the course of history, greed does not have its own special place in the brain. There are many examples of this interesting contradiction, and we will consider another one in the next chapter.

If it is true that human categories don't suggest biological structures, then one can infer a powerful principle when attempting to study emotions. A psychological phenomenon can only be scientifically assessed if the phenomenon is shown to exist in the brain. The lesson of avarice was a negative one: if one wants to make a meaningful contribution to the study of emotion, or indeed any behavior, then one may have to look past the label – even reject the label – in order to get into the neurons. But the converse is also true. Science contributes to the study of emotions by investigating those neurons where that function is represented. Remember those gargoyle statues we were talking about previously? They take on the characteristic of realism by their ability to faithfully resemble the object being depicted. If the statue is actually a fusion of several animals, we may enjoy their whimsy, but we would never hire a zoologist to search for such a creature in the wild. Instead, we would be forced to make an inference about what the real animal looks like, and then seek it instead. The same level of discrimination is needed when we seek the origins of emotions within our brains.

Looking at circadian rhythms is both the easiest and the hardest example of this kind of filtering. The SCN is one of the most important regions in the brain for studying pacemaking functions. But the pacemaking function itself is *not* an emotion. Disruption of the circadian rhythms may affect our emotions, our moods, even our mental health, but if those feelings have cerebral correlates, they will cooperate – perhaps even originate – in other areas of the brain. This means that looking at something like feeling "tired" has at least three active pockets of investigation. The first is the area where awareness is controlled, such as the circadian pacemaker. The second is the area where the inputs provided by the pacemaker are "felt", giving us our feelings of exhaustion, which may be in regions of the brain other than the pacemaker. Finally, the neurons that connect these two regions are vitally important. So in a book about emotions, day/night perceptions are relatively easy to study – we simply isolate the neurons, look at the genes, and make our correlations with emotional behavior. At the same time, day/night reactions are extraordinarily difficult to study, as the SCN is probably not the place where associated moodiness is generated, even if it is part of the cause.

While we know a fair amount about the first area, the SCN, we are almost completely in the dark about the other two. And that means experiencing yet again a strange and unsatisfying gulf, reminiscent of our other chapters on emotions and genes. It reminds me of looking through Harry's book about the gargoyles of Rouen. While I was struck by the imagination and the vitality of the sculptors, I was also frustrated, because there are satirical jokes embedded in these sculptures whose meaning and connection lie outside the art, and are thus currently beyond our grasp. Understanding that the pieces may be evident even if their underlying associations remain unknown is very characteristic of most kinds of brain research too. Considering our current understanding of the distance between a neuron and a behavior, that's not a bad way to think of emotions either.

CHAPTER SIX

Wrath

"No gloom of Hell, nor of a night allowed
No planet under its improverished sky,
The deepest dark that may be drawn by cloud;

Ever drew such a veil across my face
Nor one whose texture rasped my senses so,
As did smoke that wrapped us in that place."

-Canto XVI, The Purgatorio

"It would be a little bit like entering a city where an atomic bomb had just been detonated," the professor intoned, holding up a truly horrific 1945 photograph of the Japanese city of Hiroshima. "There would be flames and smells and smoke so thick and black you could barely see anything," he continued, *"That's* what Dante says the Third Cornice is like. It's a place filled with stinging smoke. It's acrid, corrosive, blinding, and populated with people being punished for the great sin of wrath."

We were now in the fifth day of a series of lectures devoted to the various chapters of *The Purgatorio,* the professor now explaining a much more violent place than that occupied by the lazy souls of his previous lecture. "It was no accident that Dante used the metaphor of smoke and burning to describe this level," the professor said. "Just like anger and aggression are caustic to the soul, so the smoke in this cornice is corrosive to the senses. Dante immediately wished to leave."

"Did the smoke hide the inmates in the Cornice?" the red-haired boy's hand shot upwards, even as he asked his question. I noticed that he looked crimson-faced and a bit agitated this morning, as if he had freshly come from arguing with someone.

"Of course," the professor answered. "But that's not the first thing Dante sees. He starts his journey in the Cornice on what seems like a drug trip. He hallucinates three times – visions, in his own words. Each vision is a kind of mini-morality play, complete with descriptions of instances when a person could have been vexed and wrathful and decided against it." The professor explained that the characters in the visions range from the religious, like Mother Mary and St Stephen, to secular ancient Greek rulers.

However, that was the last time things were pleasant. The professor related that as the Poets moved through the cornice, thick black smoke enveloped them. "That's when they began to hear the moaning of the inmates," he said, "They were singing some kind of unison chant, as ever, something about the meekness of the Lamb of God. That unison bit is important. In Dante's mind, wrath is the thing that breeds divisions and contentions between people. Suitable penance would be a forced unison chorus indeed." Eventually they encountered a person named Marco Lombardo, who journeyed with them to the very edge of the smoke, and the end of the cornice. Along the way, he gave a discourse on the corruption and violence of the world, whose center he said was the power and wealth of The Church. He harbored special bitterness for the then Pope.

"The journey through the smoke ends when a piercing shaft of white light penetrates the fog," the professor continued. "This is the signal that the sojourners are nearing the end of the Cornice and an angel of light is close by. Marco, who is not yet pure enough to escape the Cornice, has to leave them at this point and return to his smoky jail."

The returning of Marco is not quite the end of the poets' journey through the Cornice. Before they ascended to the next level, the professor explained that Dante once again hallucinated, seeing three more visions. These were quite terrifying, each a morality play about the aggressive destruction of unchecked wrath. The point has to do with the fleshing out of anger into aggressive action, and the great sins that are committed when human beings decide to be aggressive against each other.

What this chapter is about

As you probably suspect, the subject of this chapter is the wrath of humankind, focusing specifically on the tendency mentioned above, our tragic penchant towards aggressiveness and violence. The questions biologists might pose to Dante concern this historical and disturbing tendency of our species to become angry – even violent – with one another. Dante himself was no stranger to human wrath and destruction, as we have seen in previous chapters. But what exactly causes this aggression to stir so deeply inside us? What are the regions of the brain that mediate these destructive tendencies? Are there genes involved in the behaviors that lead people to experience angry and destructive feelings? Most of us can relate to the experience of becoming angry, and, if our life is threatened, feeling the murderous emotions welling deep within. Exactly what happens to our bodies when we become aggressive?

In this chapter, we are going to explore the biology behind the generation of aggressive feelings. We will start our journey by attempting to define exactly what "aggression" means, addressing both environmental and biological inputs. From there we will move to specific areas of the brain related to aggressive feelings, and revisit a hormone we haven't talked about since our chapter on lust: good old testosterone (though now we will behold it in its most corrosive context). Next, we will discuss the hunt for the genes of violence in mammals, ranging from mice to humans, and explain just how hard it is to define the road between nucleotide sequence and death row. At the end, we will continue our history of the mind/brain dilemmas, and, of course, add to our list of ingredients that make up human consciousness. We will ask if the presence of aggressive feelings informs the discussion of how neurons create minds, and how various hormones might contribute to the generation of consciousness.

I would like to start by describing the life of a powerful man who was no hero of Dante's: the supreme leader of the Catholic Church.

A papal definition of aggression

As stated previously, the Poets meet an inmate named Marco Lombardo on their journey through the Third Cornice. There are no historical records of a Marco Lombardo, the name simply meaning "Marco the Lombard" or "Marco of the Lombardi family". But he sings a refrain familiar to Dante, lamenting the corruption of The Church and placing blame on members of the clergy. This passion even extends to the titular head of The Church – the Pope, and Dante seems to relish special venom for Boniface VIII, who occupied the office from 1294 until 1303. Consider this passage, which is part of a solo lecture Marco gives to the Poets:

> Men therefore need restraint by law, and need
> A monarch over them who sees at least
> The towers of The True City. Laws indeed,
> there are, but who puts nations to their proof?
> No one. The shepherd who now leads mankind
> can chew the cud, but lacks the cloven hoof.

Most commentators believe that the words "chew the cud" mean spiritual reflection, and that "cloven hoof" means the ability to distinguish good from evil. It is quite a dangerous thing to say to the leader of the Catholic Church at any time, but especially about Boniface VIII, an aggressive and assertive a Pope as ever graced the throne. His life reflects both the many forms that human aggression can take and the ambiguity of trying to find a definition that would make sense to a modern-day test tube.

What do I mean by ambiguity? Consider some of the interesting ways Boniface VIII manifested his own aggressive instincts. He started his Papacy in an extraordinary fashion: by convincing the previous Pope, eighty-year-old Celestine V, to resign. He then promptly put Celestine in detention, and later condemned him to prison, where the old Pope died. Not surprisingly, this act did not endear him to members of the aristocracy and rebellions soon followed. Boniface reacted swiftly and violently. Using his papal troops, the Pope invaded rebellious territories such as the Colonna domain, and destroyed the city of Palestrina, spewing salt over the ruins. Of course, force was not the only means of political control at his disposal. As Pope, he repeatedly attempted to use the moral force of his office to control the political events of Europe, excommunicating, reinstating and then re-excommunicating monarchs and leaders who would not do his bidding. Next, he issued an edict claiming that God had decreed a hierarchy of temporal authority. The first flowed from The Church, and the second still flowed from The Church, but used secular monarchs to establish its authority. Under such philosophical freedom, Boniface's career became punctuated by the conquest of local territories. But it also had international implications in aggression against perceived enemies, this time in the form of The Crusades that he organized and financed. Finally, the Pope was accused of sexual indiscretion as well, apparently with some reason (though many of the specifics must be taken with a grain of salt, as they were uttered most loudly from his many political enemies). As we discussed in Chapter Four, he died violently, having been kept a prisoner for three days, then quickly succumbing to a fever.

The difficulty of defining aggression exactly is illustrated many ways by the life of this Warrior Pope. In his book *On Human Nature*, Edward O. Wilson divides hominid wrath into seven parts. The first he says is "the assertion of dominance within a well-organized group", well described in Boniface's ascendancy to the Papacy. A second form is "the defensive counterattacks against predators", illustrated nicely by the Pope's invasion of the Colonna domain and the razing of Palestrina. A third form is the "moralist aggression used to enforce the rules of society", which is exactly the effect Boniface hoped to achieve in his excommunications and his many Papal bulls. Fourth and fifth forms of aggression Wilson says are "the defense and conquest of territory" and "acts of aggression against prey". These tendencies are nicely illustrated in Boniface's proclivities towards general acts of warfare and his organization of The Crusades. A final form of aggression, Wilson argues, is sexual in nature, and, even though the most reliable sources from a historical perspective are the Pope's critics, the office was never far from sexual scandal during Boniface's reign.

Taken together, we have a number of tendencies from which to choose in our attempts to describe the biology of human aggression. And that's what I mean by ambiguity, because a scientific understanding of the tendency would require the isolation and study of a discrete set of parameters. Are each of these tendencies represented by a separate system of neurons coursing through our brains? Separate genes, perhaps? Maybe there are only a few systems and genes, which simply work in combination to give us our many categories. Maybe there are none, and these categories once again illustrate the futility of human attempts to organize something as complex as the behavior of the brain. From a biological perspective, what are we going to do?

It is not all hopeless

As we have encountered in each of the chapters before, finding the biology in the midst of such complexity is neither an easy straightforward effort nor a completely futile exercise. As ever, scientists look to animal models first in an attempt to gain clues as to what is occurring in human beings. In terms of aggressive tendencies, many animals display all the territoriality and violence of the aforementioned aggressive church officials. And so it is logical to begin our journey by asking two questions: (1) can we isolate the molecules that drive the behaviors in animals?; and (2) does the research have any relevance to humans? As we will see below, the answer to the first question appears to be "yes" and the answer to the second question is "possibly, but the jury is still way out".

Where then do we begin our search? Our task is familiar: we have to somehow disentangle the role of genes and neurons from the role of environment and development. While this is never an easy task, the data suggest that biology can play a powerful role in the aggressive tendencies of human beings. We will begin our discussion by describing kinds of experiments – using twins – that we have previously left out of our discussion. There is good reason for this; as a scientist used to test tubes and gene sequences, I believe the variables involved in interpreting twin studies are so numerous that scientific truth is often lost. But the data I am about to describe have stood the test of time, and in many instances point to some real biological correlates. Unlike twin studies in other subjects, it is certainly worth going over.

Where do we begin?

The most obvious place to start looking for the biology of human aggression is at the beginning: with children. We can look at two sets of kids: those that were adopted and twins (both identical and fraternal). We look at adoptions because these children dampen a variable even as they reveal a trait: if the behavior is really genetic, then the kind of home the kids are raised in will not matter. If the behavior is really environmental, then the home would matter a great deal. We look at twins – especially identical twins raised apart – for the very same reason. Identical twins have identical genes; therefore, a behavior that is purely genetic in origin would be seen in both kids, regardless of the way they were raised. Studying such children is a somewhat valiant attempt to keep the variables in the experimental design to a minimum.

So what exactly do the adoption data show? One of the most important experiments is a long-term study (twenty years and counting) of about one thousand families from Iowa. All the children studied in this experiment were separated from their parents at or near birth, and were adopted by relatives not genetically related. Some come from troubled families, with a history of criminality, substance abuse, and social malfeasance. Some do not. The researchers, under the direction of Remi Cadoret, asked a simple question: in terms of crime and violence, are the children more like their biological parents, or more like the homes in which they were adopted?

The answer to that question turns out to be a bit ambiguous, and, at the same time, clearly points a direction for future research (see Figure 18 for a summary). For those kids whose *biological* parents were not in trouble with the law and did not have any other defining social problems, how they were raised did not seem to matter. If these kids were adopted into a good home, they turned out all right. Even if they were adopted into a troubled home, these kids on average still turned out just fine. The suggestion in these data is that genes play a significant role in the social behavior of these children, and can even penetrate through the smoke of an unhappy household.

However, the story concerning children whose biological parents *were* troubled takes a whole different turn in this study. If the adopted home was stable, the kids turned out fine, just like their luckier counterparts. But such positivity begins to break down if the genetically troubled children are placed in a troubled home. The levels of aggressive behavior in criminal acts such as stealing and truancy, or noncriminal acts such as lying and school expulsions, went up 500% in this population. Such antisocial behavior appeared in the tragic placement of a child with a troubled biological background into a troubled family. In some cases, the troubled adoptee is so disruptive that he or she causes a home to disintegrate. In one particular instance, a problem four year old was adopted by a family who already had a little girl. The boy's behavior was so disruptive that the family tried to return him to the adoption agency, and only with counseling were they persuaded to take him back. They shouldn't have done it. The arguing and fighting between the parents escalated to such an extent that within two years they were divorced. The boy was eventually removed

Adoption, environment and genes

Attempts have been made to determine the role of the environment and genetic backgrounds in the prediction of criminal behavior.

AN EXPERIMENT IN IOWA

Researchers were interested in determining the factors related to aggressive behavior in adopted children. The scientists studied the fate of Iowan children whose biological parents were either socially "stable" or "unstable", as measured by a variety of factors. The results were mixed, but suggested that biological background plays an important role in the social success of the children.

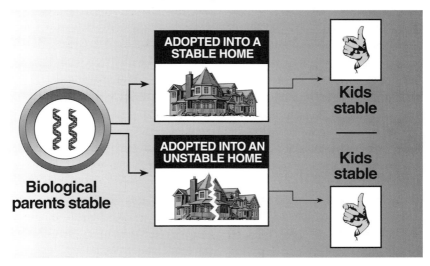

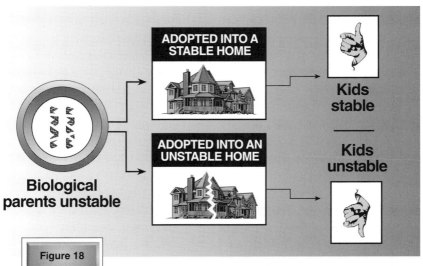

Figure 18

from the home. He responded by getting a gun and killing the daughter, a crime for which he is now incarcerated.

What does this mean?

The study mentioned above and supporting work suggest that the interaction of genes with the environment is not only extremely important, but even has predictive value. A good home can have an enormously positive effect on the outcome of a child, and a bad home a negative one. However, the story about the murdering adoptee just mentioned brings up a troubling issue, one that seems to enlarge the genetic role for aggressive behavior. Could this boy have been saved? The parents got him at four years of age. What if he had been four months? Or even four minutes?

Data about critical periods abound in the neurobiological/psychological world of human development. As anyone who has ever tried to raise a child knows, timing is *everything* which may mean that if a "good" home got a "bad" kid, the story might have a happy ending, but only if the child was rescued in time. In terms of aggressive behavior, exactly what does "in time" mean? The answer appears to come from that second set of experiments we mentioned previously, the data that come from looking at twins.

As you may know, human beings are capable of producing two types of twins: fraternal and identical. Fraternal twins are simply the product of two different eggs being fertilized by two different sperm cells at the same time. Scientists call them dizygotic (a zygote is fertilized egg) to reflect that fact. They share the same uterine environment, but are no more genetically related than normal siblings. Identical twins are a whole different story, however. Identical twins started out as a single "person", but then split into two separate zygotes. Because this splitting occurs after fertilization, the resulting people are genetically identical. Scientists call these types of twins monozygotic to reflect this fact. They not only share the same uterine environment, they share the same genes. They are truly identical.

Twins are enormously valuable to researchers for teasing out genetic and environmental issues. One can ask about predictions of behavior in one twin by looking at the presence or absence of a particular trait in the other twin (such correlations are mathematically defined and are termed "concordance rates"). For monozygotes, one can look at a disease trait such as cystic fibrosis, which is completely genetic and shows very little environmental influence. If one twin has the disease, you can guarantee with a nearly 100% concordance that the other twin will have it also. And this will occur whether the monozygotes are raised apart (separated at birth) or raised together. The reason for this strong correlation is that the disease is not environmental in origin, but rather genetic. When one compares these numbers to those for fraternal twins, you would expect *their* concordance rates to be very different (read: lower) than those of the identical twins. Why? Well, fraternal twins each have the same but independent chance of getting the cystic fibrosis gene, exactly the same as the risks experienced by siblings born at different times. Their concordance rates might be, for example, around 25%. Other diseases, such as schizophrenia, are not nearly so tight. A variety of studies have shown the correlations are more like 50% for *monozygotes*, rather than 100%. Scientists have interpreted this to mean several things, including the possibility that the origin of schizophrenia is not just genetic, but may have environmental influences as well.

So what do twin studies reveal about aggressive, juvenile delinquent-type behavior? Since males do most of the kinds of damage that social engineers are interested in stopping (more on that in a minute), much of the research was done on little boys. Looking for correlations in young boys might give us an idea about the origins of the behavior, and might even allow molecular scientists to start asking increasingly intimate biochemical questions. If we could then compare the results from studies of juveniles with the criminal behavior of adult twins, we could even start to get a timeline as to when and how genes and the environment work together to create aggressive behavior.

Just such studies were undertaken, and the answers appear to be a bit shocking (see Figure 19 for a summary). Nevertheless, they have been shown to hold up in several countries besides the United States. Aggressive behaviors appear to have both environmental and genetic influences, with one influence becoming more important than the other as the subjects get older. Consider the young boys, for example. The concordance rates between juvenile delinquent behavior in twins was found to be 91% for monozygotes and 73% for dizygotes. That simply means that if one twin were delinquent, his brother would have a whoppingly huge chance of also being delinquent. Those are very high numbers, maybe even *too* high. If the data were truly genetic, you might expect to see some fairly tight numbers for the identical twins (and you do) but vastly different numbers (read: lower) for the dizygotes. Those numbers would be just like those of our cystic fibrosis example.

Interestingly, you don't see large differences, and that caused many researchers to suspect that something other than genes was at work. It

Age and aggressive behavior

Studies have been done on aggressive behavior with identical twins in various age groups. Researchers have found that the effects of environment and genes play different roles in the aggressive behavior, depending upon the age of the twins.

Numerous experiments looking at violent behavior in identical twins have been performed. The aggregate results suggest that in younger people, the effects of environment are more important than genetic factors in determining aggressive tendencies. As the people age, however, genetic factors play an increasingly more important role (see text for full explanation).

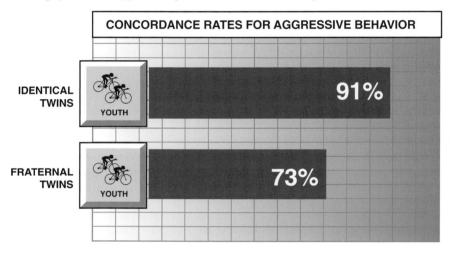

CONCORDANCE RATES FOR AGGRESSIVE BEHAVIOR

IDENTICAL TWINS — YOUTH — **91%**

FRATERNAL TWINS — YOUTH — **73%**

CONCORDANCE RATES FOR AGGRESSIVE BEHAVIOR

IDENTICAL TWINS — ADULT — **52%**

FRATERNAL TWINS — ADULT — **23%**

Figure 19

turns out that their hunch was correct, and environmental noise is the culprit. It was found that those living in a bad area of town were much more likely to be a delinquent, to the tune of about half of all kids living there getting into trouble with the law. Though perhaps intuitively obvious, the intuitions were validated with some rigor, and replicated many times. The conclusion? You could put it in the form of a question: what gene could keep you from the horrible psychological damage of living in certain inner cities? The answer is: no gene has such strength. To summarize a whole bunch of these studies, the bottom line was nothing new: environment plays a strong role in the creation of aggressive, antisocial behavior.

So that's the study of juveniles and twins. What about twin studies involving adults and aggressive behavior? They may not necessarily be living in the same hostile environment. Does one still see the same great force of the environment triggering antisocial behavior in older people? Or does something else – like the genes – begin to kick in? Of course the thing to do would be to make comparisons, like our cystic fibrosis work. If the rates for identical and fraternal twins were the same, you might once again expect that something other than genes is contributing to the numbers.

A very different pattern emerges when one examines the behaviors of adult twins, ones no longer experiencing the critical periods of development they did in their youth. The concordance rates for aggressive behavior measured in studies of identical adult twins done in several countries (and done in many different eras, including studies in the 1930s) showed the following: the average concordance rates for criminal aggressive behavior in monozygotic twins was 52% and for dizygotic twins, 23%.

Bingo. These are the differences one might be looking for if a genetic origin was truly at work. This is very different to the numbers found in youth, where the environment appears to command greater authority. With adults, the numbers suggest that genes play a greater decision-making role in aggressive behavior. Environment isn't totally tossed out (that 52% is reminiscent of the data on schizophrenia; if it were *only* genetic, then of course this number would have been 100%). However, the fact that the data from the twin studies are the same as those obtained when genes do play a role means the obvious: genes are important players in the aggressive behaviors of human beings.

So how does one look at all this stuff? If the adoption studies show us the importance of genes and environments, twin studies are dressing up two variables in interactive and surprisingly temporal clothes. Many researchers believe the following. When kids are living in the same house, that shared environment affects them deeply. If Dad and Mom are well-adjusted college professors (is there such a thing?) living in a nonviolent part of town, with the kids going to schools that aren't plagued with violence, these children will grow up to be stable and uncriminalized members of society. Conversely, if Dad and Mom are violent drug pushers living in a violent part of town, with the kids going to gang-wracked violent schools, the kids will be aggressive, criminalized, and delinquent. In other words, environment is the thing to worry about developmentally.

Once both sets of children leave their developmental environment, however, a different set of forces begins to exert controls over the person. They become less impeded – perhaps the best word here is influenced – by their developmental cues. They certainly can choose their own social structure, react to the way they grew up, essentially altering significant parts of their environment. The genes blinking on and off in their brains don't necessarily change, of course. But older adults usually have more say in the complexion of the environments in which they wish to live, options not necessarily available to them as kids. In other words, adults can help shape their future environments. The bottom line? The data suggest that as somebody gets older, the aggressive tendencies start obeying genetic papal bulls rather than strictly environmental ones. A kid might end up in "juvy" because of his parents, but, in his adult configuration, the kid will end up in the "state pen" because of his genes.

Are these data enough to give us molecules?

As stated previously, the reason I have shied away from describing twin studies in other chapters is that the data are often quite "soft", especially compared to the realm of biochemistry and Bunsen burners. Indeed, there is a sentence like this in every chapter of this book: there is a great distance between an observed behavior and the gene that causes it. Even though the adoption and twin studies just mentioned have been done, in some cases with excruciating rigor, and have even been replicated, one still cannot get away completely from uncontrolled variables. Some of these variables are important enough to affect the isolation of molecules and genes, and certainly to affect the way we view their roles in human behavior. Remember our discussion of Boniface and the definition of aggression? What does aggressive wrathful behavior in human beings really *mean*? How might our definitions affect what molecules we go after, or influence the interpretation of our data once we get some? Okay, so the twin data suggest that environment and genes play significant roles, and do so at different times. But that's an observation I might intuitively apprehend just by watching people, myself included, try to parent their children. As a scientist, my research world lies in the tiny confines of a test tube. How do we get behavior to make sense there?

To answer some of these questions, we will have to leave behind the world of questionnaires and statistical filters and get back into the world of cortexes and neurons. To stay ethical, we will even need to leave human beings altogether and find molecules in animals that appear to play roles in their aggressive tendencies. After we have answered a few questions about our terrestrial cousins, we can return to the world of human beings and attempt to organize the biology behind human aggression. Since these answers involve some fairly complex biochemistry, I would like to start our discussion not with a cell, but with an analogy involving another medieval artist. This one was named Duccio.

A lesson from Duccio and Holbein

"If I see another one of these I am just going to scream!" my friend hotly whispered, walking closely beside me in the medieval section of an art gallery. As an American, he apparently was not very used to seeing actual medieval paintings. "I mean, how much gold can one painting stand *anyway*?" We were on our way to see a sixteenth century painting by Holbein called "The Ambassadors" and, for preliminary comparisons, the tour guide had stopped in front of a medieval altar piece. This altar piece was painted by a contemporary of Dante's named Duccio, and it really did have a lot of gold leaf around it. My friend stopped talking as the tour guide cleared her throat.

"As you can see, this painting is powerful in its own right, but the characters are typically flat," the tour guide explained. "Techniques that would add realism to this work – like foreshortening and certain kinds of perspective – were not widely practised then." She drew attention to one three-paneled piece called "The Virgin and Child with Saints". She pointed at the center piece and declared, "Note the extensive use of gold leaf in this painting." I laughed inwardly and watched my friend nod with vigor.

But we weren't there to look at the use of color in medieval altar pieces. Having made her medieval point, our tour guide herded us past a number of paintings until we got to the Holbein. And it was there that my friend's interest in the subject picked up dramatically.

"This was painted more than two hundred years after the Duccio we just saw, and look at the difference!" My friend appeared a bit oggle-eyed. Here were two majestic looking young men, seeming almost as if they would pop out of the picture. Nothing flat about these characters, lots of perspective and foreshortening, and no gold leaf in sight. But the object that really drew my friend's attention lay at the bottom center of the painting. Here was a disc-shaped object that appeared to have been stretched, like it was made of rubber. It lay at forty-five degrees to the vertical of the painting, appearing to have nothing to do with anything in the composition, looking more like Dali than Early Renaissance. The tour guide sensed my friend's interest.

"Here is the real reason we are examining this picture," she declared. "What do you make of this?" and with that she pointed to the tilted disc near the bottom. Several in the group appeared to know what it was, but my friend was mystified. The tour guide addressed him. "Walk by this painting and tell me what you see," she requested. And my friend did so, and as he walked by the disc, eyes on it the whole time, an expression of wonder appeared on his face. "It's a skull!" he declared with an innocence that made some in the group laugh. "But you can only see it fully when you walk by it! How did they do *that*?"

The tour guide smiled. "It is indeed a skull. It appears to be stretched sideways, a form we now call anamorphosis. To see it fully, you cannot take it in the familiar perspective, which is straight on. Instead, to understand the disc, you have to look at it from a specific angle. From an execution perspective, that's a far-cry from the Duccio we just saw, isn't it?"

My friend was now hooked on the rest of the tour, and nodded his head vigorously in response to the guide's question. We spent the rest of the day looking at other clever examples in the evolution of perspective and realism in Western painting. At the end, my friend bought postcards of both the Duccio and the Holbein paintings and sent them to his wife.

What Holbein has to do with wrath

The realization that a concept can change depending upon the perspective of the viewer is a point I wish to underscore in our attempt to understand the genes of aggression in humans. You may recall from our study of lust that many of the emotions we experience lie just below our conscious apprehension. In this chapter, we will discuss a similar vagary, and uncover another mystery of human behavior. Emotions may simply be the after effects derived from the activity of variously stimulated brain systems. There is a convenience in this point of view, for it allows us to objectively measure emotions by investigating the underlying mechanisms of these brain systems. But there's a shadow that lies between the after effects and the neurons involved, and to understand the relationship properly we may need to change our perspective on the role of the biochemicals. We may need to do what my friend did to apprehend the skull-in-the-disc, namely take a walk around the physical data, and allow the journey to alter our perspective.

Does that sound confusing? In a few moments, we are going to examine a hormone thought greatly to be involved in aggression – the familiar steroid testosterone. We obtained initial prejudices about its function from the hormone itself by doing a Duccio, creating simple perspectives from animal studies and then projecting similar functions to humans. However, when we really looked at testosterone and human aggression, we immediately encountered an odd disc in our hypothesis. In other words, a few contradictions began to emerge. Understanding the role of the steroid in human aggression came only when we "walked around the data", changing our perspective on the nature of aggression, and gleaning a truth. Still confusing? We are going to explore all this, starting with a discussion of the areas of the brain involved in anger, and then we'll jump right into the world of testosterone and its genetic portraiture of human aggression.

The biology of aggression

To get a handle on the perspective that requires changing, we first have to understand the nature of a few of the brain systems involved in aggression. You might recall from our chapter on fear that there is a region in the brain called the amygdala. It is deeply involved in the process of fear, and has some fairly fancy sounding nerves associated with generating the experience. Of course you remember the hypothalamus, that great railroad switching area of the brain we have talked about in every chapter so far.

It turns out that both of these brain regions are involved in the experience of aggressive angers. There is a neural highway that connects the amygdala to the hypothalamus. It is called – get ready for yet another interminably big name here – the stria terminalis. As we experience feelings of anger and aggression, we find the amygdala talking to the hypothalamus via this freeway. Specifically, it sends patterned pulses of electrical stimulation to the hypothalamus, informing it of its bad mood. The hypothalamus is free to both alert and react to other parts of the brain, in an effort to inform other neural tissues about the presence of aggression. As a result, aggressive feelings are formed.

It is no accident that some of the brain regions involved in fear are also involved in the generation of anger. As you undoubtedly recall, feelings of sexual arousal also powerfully recruit the neurons of the hypothalamus. And as we shall see in a moment, hormones involved in the generation of our libidos also appear to be involved in feelings of anger and aggression. What a hotbed of activity is the area encompassing the hypothalamus and amygdala!

The role of hormones

In previous chapters, we discussed the fact that the brain regions do not generate the emotional feelings in isolation, but often use the presence of hormones to gauge their responses to a given situation. Feelings of aggression are no exception. In fact, the discovery of the hormonal sources involved in aggression are some of the oldest forms of experimental behavioral endocrinology known (as you may know, endocrinology is, among other things, the study of glands that secrete hormones into the bloodstream).

Way before Dante's time, the Ancients knew how to destroy the obnoxiously aggressive tendencies of certain animals – like bulls or rams – in their otherwise cooperative ancient flocks. They simply castrated the males. Since testicles were/are associated with sexual arousal as well, lopping off a bull's privates meant destroying the reproductive capacities of the animals in question. The conclusion? That there was a delicate association between sexual arousal and aggression; the male association between sex and wrath has been with us ever since.

This association appears to carry on with us now in the twenty-first century as well. What the Ancients were doing, of course, was destroying the source of that famously caustic male-associated chemical testosterone. (You remember from Chapter Two that testosterone can mean any one of a family of related hormones usually called androgene hormones, and that we use the word testosterone for simplicity's sake alone.) As you may know, testosterone builds up muscles – even the ones in our throats – creating huskier, more guttural voices. Testosterone makes hair sprout both where it should and shouldn't, changes the way the liver metabolizes all kinds of biochemicals, and it gets into our brains. As we know from Chapter Two, testosterone is exploited by both males and females in the generation of sexual feelings. It creates this interesting mayhem by binding to a receptor lying just below the surface of a cell membrane.

Much of the research regarding testosterone and behavior has involved studying males. In nature, it appears as though the steroid gives guys most of their aggressive tendencies, from hogging the food to fighting out the locals for mating privileges; it is thus tempting to say that testosterone causes aggression. Though the links are historically obvious, they are in fact mostly correlative. The typical modern-day experiment goes something like this:

1. Remove the source of testosterone; for example, the testicles. What happens to aggressive behavior? The levels go down, though importantly they do not cease altogether. The euphemism is the term "manageable", and when bulls are castrated they are done so to achieve just this effect.
2. Add back testosterone by direct injection into the bloodstream. What happens to aggressive behavior? Almost as if by magic, the aggressiveness returns.

That seems to make a pretty open and shut case for the association between testosterone and aggressive behavior. You can even see the effects in humans. Castrated subjects do indeed appear less aggressive. Normal subjects that have been injected with whopping amounts of testosterone appear to be whoppingly more aggressive. If one simply assays the general public, those people with higher levels of testosterone appear to be more aggressive than those with low levels. So you see the association twice. By measuring the extremes (either no testicles or too much testosterone), you find that aggressive behavior and hormone levels are directly linked. By measuring the general public, you observe that differences between individual members bring the same kind of association. So whether you are looking at animals or at humans, you appear to observe a suspicious correlation between elevated levels of testosterone and elevated levels of aggressive behavior. One might safely conclude the simple Duccion perspective, that testosterone and aggression are not only locked in a death grip with each other, but that increases in testosterone elevate levels to ensure an increase in aggressive behaviors. Right?

That conclusion turns out to be wrong

Not right. It turns out that there's a disc in the overall picture. The task in looking for causal associations between two variables – like a behavior and a molecule – is proving the so-called causal link. That is, one must not only show an association, but one must show that one variable *causes* a particular condition. In the case of testosterone and feelings of aggression, this linkage turns out to be surprisingly difficult to achieve.

We will explore the reasons for this difficulty in a moment. An alert reader, however, might at this time be wondering if we are headed towards a contradiction. As we have seen, looking at a substance and then implying a function based on the presence of the substance can sometimes bear fruit. You might be thinking of our last chapter, where conserved gene sequences in one creature imply a function when the gene is found in another creature. But sometimes the association just isn't strong enough to yield a cause. Consider, for example, an interesting passage in Dante's description of the smoky atmosphere of the Third Cornice. It mentions a zoological fact about moles. Not knowing a great deal about mammalian anatomy, I had to talk to a zoologist friend of mine to either confirm or deny what the text said. Here's the passage:

> Reader, if you have ever been closed in
> By mountain mist that left you with no eyes
> To see with, save as moles do, through the skin. . .

I read the passage to my colleague in his office.

"So is that true?" I asked him, looking around his office, eyeing some beautiful posters of giant caterpillars and butterflies he had hanging on his walls. "I thought moles were blind."

He caught me looking at the posters. "Pretty, aren't they?" he said, referring to the insects. "A lot prettier than most moles, though moles would welcome their presence for other reasons. They eat insects, you know. They're insectivores."

I was a bit impatient, "Fine. But how could they see them if they were going to eat them?" I asked again.

"The presence of an eye does not always mean the presence of vision," he responded, "And if that's what Dante's talking about, he's wrong. The kind of mole Dante would know about would probably be a European mole. Those really do have eyes – they are tiny and weak eyes – and they're hidden by fur. Even when the little guy leaves his dark burrow for lighted quarters, he still can't see." He paused. "Maybe Dante was talking about African moles. They've got eyes that are covered by skin. But they don't see either. Those eyes are vestigial."

Vestigial refers to a persisting, degenerative structure in nature. My colleague continued. "There are lots of examples of eyes that don't work in nature. There are creatures who even have eye spots, like the insects on those posters you've been looking at." He gestured to the wall. "Eye spots are mimicry on those bugs – no function at all. The insect just keeps them around. It's meant to fool predators into thinking that the bug is something other than tasty prey. See? The presence of a structure doesn't always mean the presence of a function."

And that is exactly the point I am trying to make about testosterone and aggression. After getting the address for those posters, I thanked my colleague and left his office.

Dante, like any of his contemporaries, had indeed been fooled into thinking that because moles had eyes they also had vision. A passing bird looking at those eyespots on those poster insects might be tempted to think the same thing; in fact the insect is counting on such mistakes in logic in order to survive.

While such a lack of associations might be a way to save energy or increase chances of survival, its dearth in the laboratory does not further the cause of science. There are in fact a number of ways to explain the data about testosterone and behavior. The first one is the most tempting: testosterone elevates aggressive feelings. However, there is another explanation as well: aggressive feelings elevate levels of testosterone. And still another explanation could be that neither one causes the other. It might be just like our moles, where the presence of eyes does not *a priori* mean the presence of vision, or like the eyespots on insects, where the presence of shape implies a totally different function than vision.

What is the real explanation? As ever, the distance between a behavior and a biological structure is a tenuous difficult road to follow. Happily, something like a clear picture is beginning to emerge, though it is not straightforward. Before we leave the anatomy of aggression and the role of testosterone, consider the following three experiments.

If testosterone correlates linearly with increased aggression, you might expect that addition of more testosterone would yield more aggression. Thus, measuring the amount of testosterone flowing through someone's veins should give researchers an indication of the amount of aggressive tendencies to expect. As we have stated, you can castrate an individual and watch how quickly the amount (re: frequency) of aggression diminishes.

Add back the missing testosterone and you can restore the aggressive activity, often to precastration levels. But, if you castrate someone and then add back only a tiny percentage of the precastration testosterone level, you find an amazing thing. The amount of aggressive behavior characteristic of pre-castration activity returns with a violent vengeance. You can observe the same thing if you go in the opposite direction. Castrate somebody and then add double the amount of the steroid normally seen. The amount of aggressive activity is still the same as that observed in the normals. What does that mean? It means that the brain can't distinguish between a "tiny bit" and "double the amount". And *that* means that you cannot predict levels of aggression in an individual person if their testosterone concentrations fall in those categories.

Here's another experiment. Researchers can acquire a bunch of male monkeys, put them all into the same environment and get them to socialize. The first thing the researchers find is that the monkeys form what is termed a dominance hierarchy, meaning there are vertical social rankings. The top animal can boss around anybody in the hierarchy, the second in command can do the same to anyone except the top banana, and the lower ones sort out their control issues amongst themselves (doing so in a similar linear fashion).

Now watch what happens if you inject large amounts of testosterone into one of the lower males. You might expect to incite a biochemical *coup d'état* in the hierarchy, creating a fresh dominant number one, the male with the all that new testosterone now careering around his brain. But is that what scientists observe? No. It turns out that when you add testosterone to a lower male, the male becomes more aggressive only to those males lower in the caste system than he. The injected male is still just as obedient to the higher members as before, giving up a favorite resting spot, allowing his food to be stolen and so on. He is simply nastier to the guys down the chain. The conclusion to this experiment seems straightforward. Testosterone isn't the causative agent of aggression, any more than an eye-spot gives a caterpillar vision. Rather, testosterone is simply aggravating the aggression that is already there.

This can even be observed with direct study of the brain. Remember that stria terminalis we discussed earlier? This is the region of the brain with neural fibers connecting the amygdala to the hypothalamus. When aggression is occurring, this highway is very busy, and the amygdala influences aggressive behavior by sending out strong bursts of electrical impulses along this interesting piece of neural asphalt.

The region has receptors for testosterone, which means if you flood the area with the caustic steroid, you might be able to see some changes in behavior. Like a lot, right? You are probably getting the idea that we shouldn't move so fast. And you are right, for the answer again is no. Whether you supply testosterone to the brain indirectly via the bloodstream or open up the skull and inject it directly, scientists observe the following lack of data: nothing occurs. You cannot get added testosterone to create those important pulses of electricity down the stria terminalis and activate the aggressive responses from a resting position. Such additions of testosterone can only increase aggressive behavior if the stria terminalis is *already* activated. Then testosterone can increase the rate of activity along the highway. It does this by quickening the number of impulses per unit time. And that's an extremely important point here. *Testosterone is not the origin of aggressive behavior, but rather coaxes an existing aggression into greater activity.* It is very reminiscent of those animal experiments. Testosterone is simply aggravating the aggression already there.

But what about females?

As was mentioned, most of the data we have been discussing have centered around male behaviors. There are strong reasons for this emphasis, ranging from chauvinistic masculine preoccupation to ease of experimentation. The effects of testosterone on female aggressive behavior have been studied of course. But even here, one finds a similar kind of ambiguity, and a sense that testosterone is mostly exacerbating an aggression that had been previously formed.

Consider the interesting case of hyenas. Rare in the animal world, it is the hyena female that is the stronger, more aggressive gender. It is a female hierarchy that organizes a given group, and a top female is the leader. Perhaps not surprisingly, hyena females secrete more testosterone-related biochemicals than their male counterparts. This testosterone activity changes the outward configuration of their genitals in such striking detail that it is actually fairly difficult to tell the males from the females. Once again, it is tempting to suggest that testosterone plays a powerful role in the aggressive tendencies of these animals.

Perhaps it does. The problem concerns some pesky research into hyena packs that were raised not in the wild, but in benign captivity, in fact in California. Spotted hyena colonies were "constructed" by taking infants born in the wild and raising them under artificial, namely nonAfrican, conditions. When the adult animals grew up, the females reliably exhibited the same hormone secretion profile as those in the Savannah, and with similar results. The females were larger and stronger than the males, had the same ambiguous genitalia as their wild counterparts, and possessed similar elevated levels of testosterone. The only difference between the Africans and these Californian transplants had to do with behavior. It took a long time for the females to dominate the males, such a long time that the researchers began to sense that something was amiss. Remember, these animals still had the same superabundance of androgen rioting through their bloodstream as their African colleagues. They had the same biochemical hook-ups and presumably the same overall reproductive fitness of a typical pack. That didn't seem to matter. Testosterone was not the whole story, and indeed that is the point here. What appeared to be missing was an environmental input. None of these females was able to learn about agression since there was no socially already established structure. No amount of hormone was going to jerk their behavior in the familiar way without some kind of environmental stimulus.

It is a familiar story. Understanding these data requires a shift in our perspective, for the face of it seems to point to testosterone's being a cause. Instead, we kind of have to walk around the data to see a truer image, much like that Holbein painting with the skull-in-the-disc mentioned earlier. Testosterone might be able to aggravate an aggressive tendency already in progress, but it does not appear to create the original.

That little walk not only yields a more proper perspective, it also yields a more informed question. The steroid only exacerbates aggressive processes that are already in motion, all well and good. But then getting at the biological correlates of aggression necessarily means looking deeper than testosterone, and getting into those processes the steroid aggravates. What is the nature of those processes? And what genetic/environmental triggers cause them to create aggression. Most importantly, how do we relate what we see in the neuron with those adoption and twin studies mentioned at the beginning of this chapter?

Obviously, we need to look elsewhere in our attempts to understand the molecules behind the brain systems of human aggression. It is not enough that we know a lot about the cells and brain regions involved in our wrath. We are familiarly forced to look under the hood of the neurons that govern these processes, which means peering into the world of the gene, to know more. Such a journey always changes our perspective, and understanding the DNA behind aggression is no exception.

If testosterone is not necessarily causative, scientists have to ask new questions. Here are two of them: (1) are there other systems to look at? and (2) do these new systems, if they exist, apply to humans? The answer to the first question happily is "yes", and we will discuss two interesting gene systems that portend great research promise for the future. The answer to the second question will sound familiar to you: we don't really know if the data can apply greatly to humans. Like the twin studies, the jury is still way out. There is a great deal of controversy surrounding the notion that much of our violence is written in our genes.

The first gene system we will discuss has such an odd biology that an analogy is required to explain its function. The analogy will take the form of story that happened back when I was a little boy.

The ghost in the machine

I was born into an American military family and, as a result, lived a great deal of my life in Germany. This story happened in an area of Germany that had many rural castles, all of which lay in ruins.

My family was close friends with a number of other families and we would often take camping trips together. On one outing, with one of those families, we made camp not far from one of those old medieval castles and while we roasted marshmallows in the cool of the evening, my father decided to tell us kids a ghost story. Our backs to the tents, facing the fire in the shadow of the ruins, it was not hard to create a spooky atmosphere. My father was a master story teller and soon we were all enthralled in a strange tale about ghostly missing limbs of ancient knights. At the climax of the story, my father whispered that the phantom arm of one unfortunate knight haunted the very woods in which we were camped, and it was known to strangle little boys and girls in the dark of the night. Immediately my Dad cried "Look behind you!" and as we whirled around, an arm in a medieval gauntlet appeared to emerge right through the wall of the tent behind us. "Aaieee," I remember shouting, and took my hot marshmallow stick and jammed it in the general direction of the arm. The stick missed the arm, and instead went through the tent slit. I immediately heard an angry, cursing adult voice. Amongst the phrases I could repeat here were such things as "Ouch, you little brat!", and I instantly recognized the voice of one of my father's friends. The conspirator unmasked, the adult clambered out of the tent, throwing down his toystore glove, a hot gooey marshmallow stuck to one side of his face. My dad laughed so hard he could not finish his story. But I wasn't laughing. For all I knew, a medieval arm had magically penetrated the tent and was going to wreak havoc on us until I saved the day.

We did not go out camping with that family ever again.

What this has to do with genes

There is no doubt in my mind that the aggression brain subsystems we have been talking about were alive and active in my father's friend that evening – especially targeted at my dad and his youngest male offspring. But I do not tell this story of penetrating limbs to illustrate the violence of American military men in the early 1970s. Rather, I want to use the idea of ghostly penetrations to relate the biology of one of the most mysterious molecules ever to activate a gene. The molecule is a gas called nitric oxide, and has so many powerful functions it was once named the molecule of the year.

Because nitric oxide is a gas, it works a lot like that phantom ghost arm in my father's story. Nitric oxide can penetrate any tissue or cell once it has been made – and it does so by literally diffusing out from its point of origin. There is no need for it to be escorted away from the cell of origin, nor is there any need for it to lock onto some protein on the surface in order to exert its effects. Once made, nitric oxide works like a ghost, simply floating out from the site of manufacture, drifting into a target cell, turning on genes once inside. As we'll see, some of these genes mediate anger as real as my father's ex-friend's.

The where-and-when decisions of nitric oxide's manufacture are very important because of this ghostly property. Once the nitric oxide is made, there is very little control over where it floats. There is a gene that encodes a protein that can make nitric oxide, appropriately termed nitric oxide synthase (NOS). Not surprisingly, the activation of this NOS gene means that the cell (1) will be capable of making the ghost and (2) that the surrounding cells will be immediately affected once the ghost has penetrated the cell of its origin. Humans have several kinds of NOS genes, including one that is active in neural cells. It is the gene for this neural form, called nNOS, that is relevant to our chapter on aggression.

What NOS has to do with violence

Even though humans have nNOS in their bodies, the experiments I am about to describe were carried out on mice. To describe what I mean by all this, we need to examine some background biochemistry and then talk about some very interesting data on genes that do not appear to involve testosterone, yet are fully capable of creating aggressive responses.

The presence of nitric oxide in mammals and the genes that synthesize it opened up a world few thought existed a decade ago. So profound is the depth of its action and the breadth of tissues involved, that the only way to adequately explain it is to make a list. That list is shown in graphic form in Figure 20.

As we discussed, nNOS has been found in neurons, and is involved in functions ranging from processing the things we smell to the perception of pain. Nitric oxide can also serve as a neurotransmitter, as well as a growth factor to which neurons are attracted.

Most importantly, *nNOS* gene expression has also been found in specific areas of the brain known to be involved in aggression. These areas have been termed the limbic system. Deep within our brains lies a connected ring of structures surrounding our olfactory bulb, the place where we process smells. This clustered mass of neurons was named the limbic system in 1952 by Yale neurobiologist Paul MacLean. The limbic system is composed of many parts, including two of the most familiar to you, the amygdala and the hippocampus (the hippocampus, as you recall, is deeply involved in the creation and storage of memory). Stimulating various regions of the limbic system can result in a wide variety of behaviors ranging from pleasure and fear to rage. The fact that the *nNOS* gene was specifically expressed in brain regions known to cause rage made scientists become very interested in its role in aggression.

The technical knock-out

As we have discussed in Chapter Four, one of the most powerful and elegant techniques for elucidating the role a gene plays in the behavior of an animal is also a selectively destructive one. Protocols have been created that allow scientists to obliterate one gene – and one gene only – in a developing embryo. By the time it is born, the animal will contain all of its normal genetic information except the gene that has been disrupted. You may remember that we call such animals "knock-outs". The roles of behavior and of the gene can be assessed by simply comparing the knock-out mice with identical twin controls that have not been genetically manipulated.

This technology is so specific that it can be applied to this *nNOS* gene,

The functions of nitric oxide

Nitric oxide plays an astonishing number of physiological roles in the human body. Here are four of the most prominent.

THE BRAIN
Regulates release of pituitary hormones; regulates the sleep cycle; involved in the formation of human memory; mediates dependency and tolerance of many drugs.

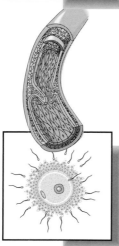

THE BLOOD AND CIRCULATION
Regulates blood pressure. Helps blood cells in the defense against pathogenic viruses, protozoa and certain bacteria.

DEVELOPMENT
Regulates the growth of certain skin cells, immune system cells, and skeletal muscles in the developing human embryo.

CELLULAR ACTIVITIES
Regulates the activation of specific gene sequences; helps control translation of mRNA; involved in "cellular suicide" (apoptosis); helps regulate cellular energy supplies and waste disposal.

Figure 20

the one whose protein was found in the limbic system, without disrupting any of the other *NOS* genes found throughout the body. And this is exactly what a group of researchers did. The embryos were created, the technique applied and *nNOS*-minus mutant mice were born soon after.

What the researchers did next

Admittedly, the experiment was something of a crapshoot. Since nitric oxide does so much in the human body, exactly what behavior was going to be uncovered was not known. And to the researchers initial disappointment, the mice looked normal enough when born and were allowed to grow to sexual maturity. For better or worse, however, this growth into adulthood is when all hell broke loose.

That there might be something very odd about the behavior of the knock-out mice was first surmised by technicians who cleaned their cages. To ensure a consistent environment, control mice that were not manipulated were always kept in the cage along with the knock-outs. And every morning, the technicians found that one or two of the littermate controls had died, and that the manipulated mice were still alive. The way the controls had died was quite disturbing, for it appeared that they had been savagely attacked. When more control mice were placed in the cages, the technicians invariably found them dead next morning. These deaths only occurred in the cages where the knock-outs were male (with female knock-outs, no such grisly observations were *ever* observed). The connection was obvious and a hypothesis was made. It seemed that there was a direct correlation between these deaths and the *nNOS* male knock-outs.

The scientists of course placed the knock-outs under twenty-four-hour observation. What they saw under simple observation was astonishing. Within ten minutes of introducing a knock-out male to nonknock-out male counterparts, the genetically manipulated mouse was attacking the control. A horrible, violent, aggressive fight ensued, inevitably resulting in death. That explained the results the technicians were reporting. If two knock-out males were introduced into the cage at the same time, the violence commenced in less than one minute.

The researchers decided to quantify this behavior using what is called the intruder-resident model (a stricter variation of the observational data described above). These more formal observations confirmed what everybody in the laboratory suspected. Offensive behavior on the part of the knock-outs was immediate, violent and lethal. Direct attack, biting, wrestling and chasing were all observed and scored as a function of time. The aggressive mice almost never exhibited any submissive behavior (rolling onto the back with paws extended) to any of the other males in the cage. In contrast, the controls did so constantly, yet received no mercy or relief from the aggressive animals. Eventually the researchers stopped the experiments fifteen minutes after introducing the two sets of mice to one another because serious wounding and death were so often the consequence.

What does that mean?

The data linking aggressive behavior to nNOS is fairly strong in this study. The manipulated mice were tested for differences in strength and agility when compared to controls, and there were no differences. They scored the same as controls in an open-field test, a protocol that measures normal levels of anxiety. It's just that the knock-out animals were more aggressive. A genetic link between an important series of behaviors was established. Mice with normal nNOS don't do this.

There are of course two questions: (1) is it *really* the causative agent, or just one link in a chain of molecules, or ancillary, like our testosterone studies? and (2) can it be applied to humans? The answer to both is: we have to wait and see. The data do nothing except tell us about mice and their behaviors when somebody wipes out their *nNOS* genes. In addition, these mice seem to show similar testosterone dependencies. For example, if one castrates these aggressive mice, they quickly settle down, exhibiting none of their former murderous behaviors. If testosterone is then added back to these castrated mice, they resume their former, extremely aggressive ways. Sound familiar? You bet, if you recall our Holbien metaphor, which simply means that our need to shift perspectives of the origin of aggression was well-founded.

There are many caveats to the applications of these data – especially to humans – and the usual suspects of species problems, environmental background noise, behavioral categorizations, etc., etc. apply. As mentioned, nNOS may simply be one link in a chain of molecules involved in the creation of aggressive mice. Of course even the weakest link in a chain is enough to untether a gene from a behavior, and nNOS may not even be the most important sequence to understanding rodent aggression.

Don't get me wrong as I mention these caveats; there is a great deal of value in this work. But with all of these cautions mentioned, the data are mostly important for a single reason. The information gives researchers another flashlight to shine on humans, a new gene sequence to consider in attempting to paint the picture of human aggression. As nNOS biology does not appear to be directly related to testosterone – indeed the testosterone system seems to be preserved in these animals – we may have our first glimpse into the mechanism that generates the behavior in the first place. At the very least we may have before us one component in a complex interlocking series of molecules that work together to create wrath and aggression in mammals.

Another system on the horizon

The idea that more than one biochemical may be involved in certain forms of aggression is strengthened by research involving animals a little closer to home: monkeys. This involves the study of a class of biochemicals we have previously discussed, a class termed neurotransmitters. As you recall, neurotransmitters are chemical messengers two neurons use to exchange information with each other. One neurons "throws" out the neurotransmitter, and the other neuron "catches" it. This catching involves the presence of protein molecules we have called receptors. When a neurotransmitter sent out by neuron A binds to a receptor found on neuron B, information is transferred.

One famous neurotransmitter is serotonin. Different neurons use serotonin to transfer specific pieces of information. Serotonin is used by the body in so many different mood-affecting processes that it was a natural candidate for study in the attempt to understand aggression. Experiments very similar to the work with testosterone were performed on animals using artificially elevated levels of serotonin. But the results achieved were very different to the steroid data. If serotonin levels in monkeys were artificially elevated, the animals actually became *less* aggressive. The same thing was found in mice and rats. If serotonin levels were lowered, the animals became more impulsive and more aggressive.

Correlative data concerning serotonin and aggression were also found in primates living in nonlaboratory conditions. An interesting and quite wild colony of rhesus monkeys lives on Morgan Island, a speck of land off the coast of South Carolina. Because they are truly "living in the wild", these monkeys make ideal subjects for experiments in behavior research.

Aggression is one such behavior that has been examined in this colony of primates. This was done in a formal fashion by measuring the number of scars and old battle wounds a given animal possessed, and also by live recorded observations of animals in their territory ranges. While not perfect measures, researchers used these behavior traits to make judgments as to the individual aggressive behaviors of members in a given troop.

In one study, male monkeys were divided into groups and their aggressive behaviors assessed. Some of the primate members were really very feisty, picking and receiving fights and battles with sometimes alarming frequency. Not surprisingly, these males carried many of the souvenirs of their encounters, with scars on their bodies. Other members appeared more docile and passive. These males also carried fewer reminders of socially hostile encounters on their skins. Both sets of males underwent spinal taps and their serotonin levels were assessed from the tapped fluids. Consistent with the addition experiments described previously, the most aggressive males had the lowest amount of serotonin sloshing around their brains. The least aggressive males had the highest amount of serotonin.

To make this correlative data into something more causal, genetic intervention experiments would have to be performed. Some of the best are those knock-out experiments we discussed in the nitric oxide story above. Researchers knew that the data from monkeys were also available from mice, and, since current technology for mammals uses mice, the experiments became obvious. What if you disrupt the effects of serotonin by, say, knocking out one of the receptors to which serotonin normally binds? That way, even if serotonin were secreted into the space between two neurons, it would have nowhere to dock. Since the effects of serotonin depend on receptor interactions, such knock-outs would be quite valuable indeed. Researchers decided to initiate such experiments, using a serotonin receptor going by the name of 5-HT1B.

And what did the researchers find? The mice that suffered the mutation were turned into aggressive berserkers, just like those nNOS knock-outs. The incidence of violence went up 600% when compared to controls. It was easy to interpret these experiments in the light of the previous work. Serotonin levels could be interpreted to play a role in dampening the aggressive tendencies of male animals. When little serotonin was around, the animals could become quite aggressive. Knocking-out the receptors in the mice was just like dropping the serotonin levels; by disabling part of the "peacemaking" apparatus, one was ensuring the presence of aggressive behavior.

So is serotonin the magic bullet, the great organizer of aggressive behavior which testosterone can only aggravate? The answer turns out to be no, at least in primates. You can actually do a pretty cruel experiment with a group of monkeys. They always sort themselves into their dominant hierarchy, but this order can be artificially reorganized (by direct human intervention) and the least dominant males will become the most dominant, and the most dominant males will scrape the bottom. When the researchers checked the serotonin levels of the former dominant males now forced to the bottom, they found that their serotonin levels had changed. Changes in serotonin were also noted in the newly "crowned" males, the ones that had formerly been on the bottom. The conclusion? Serotonin levels can actually be manipulated not by changing the animal's biochemistry but by simply altering the social position. Rather than being genetically hardwired, levels of the biochemical turn out to be quite flexible indeed. Serotonin biology does not appear to be the original all-consuming source of aggression after all.

What can be applied to human beings?

Taking all these data together, what can we say about the genes for aggression in human beings? We all want the easy explanation, that specific biochemical differences lead to specific differences in human behavior. As we have seen when considering other behaviors, such hopes are futile. Human behavior, even in this important issue, is enormously complex and intertwined with unsuspected – and in many cases unknown – variables.

That hasn't stopped researchers from attempting to go after genes involved in human aggression; in many cases their efforts have drawn a great deal of wrath and criticism from certain members of society. Let me relate to you a real-live incident that occurred when a group of scientists attempted to address the subject of genes and criminality at a formal scientific meeting.

The setting was peaceful enough, a retreat center in the middle of rural Maryland. The meeting was supposed to be about genetics and the roots of human violence. But as the lectures began, the auditorium suddenly erupted into mayhem. Bursting through an unsecured rear door, angry protesters exploded into the auditorium, shouting and screaming slogans about racism and genocide, waving red flags, thoroughly disrupting the conference. Most of the attendees were in shock, a few scientists shouted back, one even got violent (!) with a member of the invaders.

Why would such work be so controversial that a meeting about facts would become incendiary? The implications of genes and behaviors are always potentially damaging, especially when one considers that twentieth century humanity did not always do well with genetic ideas. Take the United

States in the early 1930s. With the country in the ugly grip of the Great Depression, the crime rate became much greater than it is today. Unfortunately, such misery coincided with a social fad known as the eugenics movement, a set of ideas in which genetic inheritances are thought to play major roles in human personality. One of the tenets of the movement was that the human race could be purified by allowing only those "fit" individuals the right to procreate. The ideas of eugenics led to the forced sterilization of thousands of so-called "feeble-minded", "mentally ill" – and here's the kicker – "habitually criminal" individuals. It reached its nadir in Nazi Germany, as you well know, where genetic inferiority was the philosophical excuse to gas millions. That's scary stuff, and there is a great fear amongst many professionals that the techniques of genetic engineering could herald a new age of eugenics. In the Maryland meeting, it didn't matter that the protesters were confusing individual differences with group variations. Even the hint that we might march towards genetic identifiers of violence was enough to incite the violence.

Wishing to avoid the great thickets of human prejudice does not make aggression go away, however, and there *are* a few things you can say about it genetically. One of the most striking observations about human beings and aggression has to do with the gender that does the damage: most human violence is carried out by males. In the United States, men commit orders of magnitude more rapes than women commit. Men are indicted five times as often as women for aggravated assault, and commit ten times the number of murders. The imbalance holds true even if you look at males from a variety of cultures around the world. Men commit more murders than women in Europe, more in Africa, more in India, in fact everywhere.

With such extreme numbers, it was natural for scientists to begin looking at especially violent males in an attempt to understand aggressive behaviors in humans. One of the most famous missteps occurred in the late 1960s, with the examination of the number of Y chromosomes in certain prison inmates. As you remember from high school biology, the Y chromosome determines the presence of the male gender in the human population. The idea to look for extras went as follows: if males did most of the violence, and all males carried the Y chromosome, maybe a mutation in the Y chromosome was the reason for the damage. Or, in keeping with the linear nature of testosterone experiments, maybe more Y chromosomes meant more aggressive behaviors. It must be said that the presence of extra Y chromosomes was not necessarily far-fetched. It is true that a small percentage of the male population carries more than one copy. But more *violence*? The idea at least led to a testable hypothesis. Why not look in the places where the violence had been proven – like in prison populations – and assess the number of chromosomes in the most violent inmates? If a correlation could be made between aggression and an extra load of male DNA, one might come up with a gene.

The initial results of the researchers confirmed their suspicions and made for some rather unfortunate headlines around the world. It appeared that, in certain prison populations, there were five times as many men with the extra Y chromosome than in the general population (and in one study, nineteen times as many of the "super-males"). This caused an immediate sensation, and certain hospitals even started screening their male babies for the extra genetic baggage.

It turns out that the excitement was premature. After some rather unfortunate twists and turns, a powerful study was performed in Denmark looking at thirty thousand different men, attempting to replicate the initial findings. They couldn't. In their hands, they found a surprising lack of "super-males", even though there was plenty of violence to go around. The only association the researchers found was that males with extra Y chromosomes tended to be less intelligent, as measured by a series of mental exams.

These findings have held true to this day. Now we know that the double Y chromosome is so rare, and male-only violence so common, that the extra DNA cannot be a major cause of human violence. If "super-males" are prone to crime, the phenomenon may be due to the indirect effect of intelligence on behavior, rather than any tendency towards aggression.

This setback in the attempt to understand the biology behind human aggression was only temporary. As more data from the animal world suggested strong biological components in aggressive behavior, the temptation to look for physical explanations in humans remained great. In 1979, researchers began looking at serotonin levels in humans, specifically focusing on certain men in the US Marine Corps. They found that those soldiers who were prone to excessive violence, even psychopathic deviance, had something in common with each other biochemically. When the researchers did spinal taps on so-called problem marines, they found the soldiers all had decreased levels of a chemical called 5-HIAA (mercifully short for 5-hydroxyindoleacetic acid). That got the researchers fairly excited, because 5-HIAA is a breakdown product of good old serotonin. Was it possible that these violent men had less serotonin than nonviolent men? The same lowered levels were found in a whole range of society's most troubled citizens, ranging from children who tortured animals to career criminals. The dramatic nature of this work led to experiments on monkeys and other laboratory animals, work we described at the beginning of this chapter.

Since that time, a number of efforts have focused on a laundry-list of biochemicals and physiological processes in an attempt to describe the biological basis of human aggression. Genes with names like monoamine oxidase A (an enzyme that naturally breaks down serotonin and is found in a family with a powerful history of criminal activity), physiological phenomena such as low resting heart rates (seen in children prone to violence), even elevated levels of the metal manganese in violent offenders have all had their turn at being the one and future biological explanation for aggressive human behaviors.

Can we make sense of all these data?

It is very tempting to put together simple explanations for a tendency that has had such a powerful – and often horrible – influence on human history. Unfortunately, and as I have stated several times, the jury is still out regarding the specific biological origins of human anger and aggression. Anatomy has shown us that there are specific brain neurons involved in the feelings. But that's not surprising; the brain is ultimately responsible for *any* behavior available to the human experience. Testosterone has mostly shown us that hormonal explanations of human violence are not straightforward, and that the primary aggression brain systems at the molecular level have yet to be uncovered.

The search to find these "missing" components have not been very fruitful either. Nobody really knows about nNOS in humans (the experiments have yet to be done), the XYY hypothesis has been thrown out, and there are ambiguities regarding the roles of serotonin and steroids. Explanations involving other candidate genes, molecules or physiological processes either have not borne fruit or are too premature to say anything about responsibility.

Which leads us to what, besides a great deal of confusion? It would be one perspective to say "not much". Indeed, considering the number of false leads concerning the genetics of human violence, we have many more questions than answers. But there is another way of looking at these data, a different perspective which is both exciting and encouraging. We did not always think of aggression in terms of a problem of genes interacting with environments. We used to think of it exclusively in terms of evil spirits, or the whimsies of planets spewing forth their astrological determinisms.

However, now we are taking seriously the roles of physics and chemistry in living tissues, and the way that social forces interact with all of them in human behaviors. We are briefly reviewing the history of mind/brain debates and trying to define human consciousness partly because we want to show the revolutionary nature of the scientific perspective.

In the last part of this chapter, I would like to use this optimistic biological position to return to our discussion of the history of minds and brains, as well as introduce yet another ingredient in the definition of human consciousness. We shall end the chapter by re-examining the notion of perspective, considering why studies of emotions like aggression can give us information about their biology, and how their violent existence hints to the biological substrate from which they spring.

Back to minds and brains

In the last chapter, we talked about the first glimmerings of an extraordinary idea: that issues of the mind might be understood in physical terms, in other words, that the mind might *be* the brain. A number of philosophers came to the conclusion that the soul, like a sophisticated clock, might simply be an enlightened machine. These ideas found their roots in a reaction to the increasingly absurd solutions people were offering to solve the mind/brain dilemma. You might recall that Berkeley believed that only internal perceptions of mind drove reality, and that Leibniz suggested that minds and brains coexisted but never interacted. By trying to solve Descarte's problem of dualism, the seeds were laid for its destruction.

The next great events in addressing the mind/brain dilemma led to many significant milestones, and a few pesky detours. The most significant milestone had its physical roots in materialism: if the brain could act like a clock, then its various components might be isolated and reconstructed, just like those of an intricate machine. This philosophical assent gave researchers permission to explore the brain using the tools of the physical sciences. The discoveries made by various researchers who capitalized on such notions revolutionized the question: Pierre Broca discovered the speech center in the brain, and Camille Golgi invented a silver stain that makes neurons pop-out under the microscope like piles of twigs. The great neurobiologist Santiago Ramón y Cajal discovered, among many other things, that neurons are separated by tiny gaps. Sir Charles Sherrington detailed how reflexes work by showing how the body physically interacts with the brain – *physically* interacts, mind you. We were coming a long way indeed.

Even that thorny question of theological quandary, the existence of human mental illness, began to find illumination once scientists had intellectual permission to explore the question from a physical point of view. One of the greatest scientists to ever explore the question of mental illness was Emil Kraepelin, widely hailed as the father of modern biological psychiatry. He was inspired by the work of people like Broca, and assembled one of the most gifted teams of researchers that has ever graced a university (including Alois Alzheimer, the famous discoverer of the disease that bears his name). Kraepelin's major contribution to psychiatry was to describe and identify specific types of mental disorders as *illnesses* – just like whooping cough, or tuberculosis. He noticed that people who had mental disorders might get better quickly, deteriorate quickly or experience periodic bouts with symptoms, just as if they were suffering from a disease. This portended a great deal of potential hope; if the brain was indeed like a clock, then it might be fixable were it ever to break. So, mental illness might just be caused by a broken machine in the brain, rather than an evil spirit.

The question has enough variables that a few detours were bound to occur. Other researchers contemporary with Kraepelin were not interested in the architecture of the *brain* as fleshed out in neurons and dendritic trees and synapses *per se*. They were interested in the architecture of the *mind*, and sought to understand how subjective emotions and personal experiences might result in specific pathologies, even present behaviors. I am of course talking about researchers such as Sigmund Freud, Alfred Adler and others of the so-called psychoanalytic school. I use the word detours because most of what Freud postulated turned out not to be true. He was not a charlatan; indeed, he pioneered ideas about human nature that inspired powerful research and lay the groundwork for some of the best research done in human biology. Instead, by attempting to map the boundaries of the human mind, rather than asking how neurons turn into behaviors, he made both valuable and distracting intellectual forays into the human psyche. At once there were two different points of view – the neural and the psychological – and each was making the claims of science on their individual insights.

The major contribution of the biological researchers to the mind/brain debate was to consolidate the gains of the materialists we discussed in the last chapter, and begin to flesh out their accomplishments in terms of cells and biochemicals. The major contribution of the psychological researchers was to start at the other end of the question – the mind end – and work towards the same goal by using psychological (and what turned out to be theoretical) constructs. We have done a very similar thing in our attempt to understand the notion of human aggression. We looked first at mind-like issues, using twins, asking questions about environmental influences in the creation of disturbed individuals. We then went to the other end of the question and started over with genes and cells, talking about nNOS and serotonin and the possibility of chromosomal super-males. In many ways, the modern way we look at mind/brain issues ping pongs between the precedents these two German giants, Kraepelin and Freud, set for us at the beginning of the twentieth century.

Another ingredient for consciousness

The twentieth century was quite remarkable, and in it a great deal of valuable physical data became available describing in detail how parts of the *mind* function. With all this exciting work floating around, it was natural for researchers to begin exploring that last bastion of the mind/brain debate, human consciousness. Our discussion throughout this book on consciousness details the specific accomplishments of some of these twentieth century investigative heroes. We find they discovered that the notion of consciousness is a complicated mixture of processes, neurons, and ideas. There are the notions of long-term memory, of short-term buffers, of working memory, and even of certain arousal systems.

There is an important ingredient uncovered by twentieth century researchers that we have left out of our discussion of consciousness, even though we addressed it earlier. In Chapter Two, on lust, we found that many kinds of emotional processing occur in the absence of conscious awareness. Most researchers agree that these unconscious processes make an enormous contribution to our feelings of awareness. Let me explain more clearly how this applies to a definition of consciousness.

As we have discussed, you can experience an emotional feeling and not be consciously aware of why you experienced it. But that does not mean the stimulus was refused a cerebral entrance or, critically, that it did not contribute to our conscious feelings. Some signal may not be consciously noticed yet still trigger powerful responses in the human body and brain and, ultimately, awareness. In that case, the content of the stimulus may enter working memory through some other route, as a reaction to those powerful body or brain responses, and then burble into our conscious awareness.

Psychiatrists run into this kind of backdoor phenomenon all the time, and I include it in our discussion on wrath because the emotion illuminates it particularly well. A woman may be humming songs like *"You Ain't Nothin' But a Hound Dog"* when thinking of an old boyfriend who dumped her. A person may continually forget to go to meetings that they would rather not attend. The phenomenon even shows up in the therapist/patient relationship. It is quite common for patients to begin treating their therapists like they treated previous authority figures, for example. This can be quite negative if the authority figure was resented, or quite positive if the therapist brings to mind someone very special. The phenomenon is so common that it has a name. Psychiatrists call this unconscious intrusion into conscious awareness transference.

The upshot? A person may consciously experience emotions, but not know where they came from. As a corollary to the previously presented idea that emotions are things that *happen* to us, rather than things we *will* to occur, part of the shape of conscious experience comes from underlying unconscious stimulations. The point is almost counterintuitive, yet it is a fact: unconscious stimulations are a very important component in the list of ingredients that comprise consciousness.

Concluding thoughts

Here we are then, thinking about aggression, minds, brains, emotions, and consciousness, and using science to address their boundaries and understand the relevant neurons. That's quite a leap, considering our history; what started as a soul in a fleshly bottle has ended up being genes in an environmentally impacted brain. How could we do such a thing?

By studying the biology behind aggressive feelings, we begin to see through the thick black smoke of ages past, and uncover a valuable truth in our exploration of human behaviors. While we are not yet certain what aggression system hormones like testosterone are affecting, we know that the hostile feelings and responses are after-affects caused by the system's activity. This linkage gives us permission to study emotional responses objectively and, in so doing, illuminate the system for generating the conscious feelings. And that's a first. Our study of wrath shows us how we might link a conscious experience to the activated emotional system under study. A bridge, if you will, is in sight between the two parts of LeDoux's model of emotional experience, the subsystem being activated and the awareness of its stimulation.

We even get some help from laboratory animals, as long as we keep our intellectual noses clean. Since the brain systems that make our emotional responses appear to be very similar in animals and people, we can use animal data like a filter, asking and answering important questions about similarities and differences. That's a far cry from concepts such as an evil spirit, or even something like Freud's Superego. Similar to that odd disc in the Holbein painting, we now understand that the data were there all along, but we had to change our perspective on the question in order to see them properly. Understanding that real live brain systems stand behind our generated feelings, even if we are not always consciously aware of them, is not a bad way to look at human aggression. Indeed, it's not a bad way to look at most of the behaviors mentioned in this book.

CHAPTER SEVEN

Envy

"Just as the sun does not reach to their sight,
So to those shades of which I spoke just now
God's rays refuse to offer their delight

For each soul has its eyelids pierced and sewn
With iron wires, as men sew new-caught falcons,
Sealing their eyes to make them settle down."

-*Canto XIII*, The Purgatorio

"How would you like to go through life with your eyes wired shut?", the professor began ominously, instantly commanding the attention of the class. "And I mean *wired*, like somebody took a heavy duty needle and sewed them closed with metal thread. The reference Dante uses has to do with falconry. But we are not playing sport here. Dante and Virgil have just met with yet another company of miserable people. They are walking into the Second Cornice, reserved for those enslaved by the great sin of envy."

We sat still for a minute. I knew from falconry that diurnal birds usually stay motionless in the dark. That's why they use black hoods. In the old days, the trainers would first sew the eyes of newly caught birds closed as part of their training for hunting. Only later would the threads be cut and the use of black hoods permanently employed. But why would such a practice be the punishment for envious souls? The same question seemed to be going through the red-haired boy's mind. He asked "What's with the metal wires?"

The professor cleared his throat. "It is a form of reversible blindness," he said. "In Dante's world, envy is an offense of the eyes, and so the eyes are wired shut as punishment. Even the color of the rock from which the cornice is made – purple – is the color of a bruise. The idea is that whenever somebody had good fortune on Earth, it bruised the world of the envious in purgatory."

As with avarice, this notion of material wealth arousing feelings of envy was not far-fetched to the people of fourteenth century Italy. I remember the professor saying that Florence was increasingly wealthy in Dante's time, establishing trade networks from London to Cyprus. That meant they used a lot of oceangoing ports to ply their goods. One of the most popular Florentine goods in those ports was textiles.

"Dante even used the clothing the inmates wore as a form of punishment," the professor continued, referencing his talk on Florence's use of the sea. "All the inmates in the Second Cornice wore cloaks of haircloth. That's the clothing of the penitent, a stiff coarse fiber made from goat hair. Such a reference would not be lost on the Florentine population."

It was instantly apparent that the Second Cornice would be no friendlier a place than any of the other regions of Purgatory. The shades in the Cornice were talkative however, and Dante recognized a number of people as they sojourned through the level. They met a woman named Sapia, from the town of Siena, who had some interesting things to say about her life of envy, as well as the history of her home town. It is no accident that Florentine Dante mentions Sapia or Siena, because their respective cities engaged in a conflict crowned by the Battle of Montaperti. It was a conflict that affected the future of Florence – and Dante – in a very profound way.

"He actually knew a few other people in the Second Cornice besides Sapia," the professor concluded, "including several mysterious figures whose identities are lost to history, that is if they are not wholly fiction. Dante records his conversations and then moves on hurriedly. As you might expect, the Second Cornice is a spooky place, and they hear towards the end two bodiless voices – ghost-like – talking of the consequences of the sin of envy." Their tour is concluded with a denunciation (spoken by Virgil) of human stubbornness and a refusal to see the glory of the heavens, and they encounter another angel before leaving.

What this chapter is about

You might predict from these spooky introductions to the Second Cornice that this chapter is going to be about this visual sin of envy. I have to admit to a certain amount of frustration in considering how to tackle the topic. I want very much to address the feelings of jealousy in strict biological terms as we seek to understand the biology of human behavior. But I immediately run into a fairly insurmountable problem. How does one discuss the *biology* of envy? The reason there is an insurmountable problem is that no one has ever isolated a gene responsible for human jealousy. No one has ever identified a region of the brain that is exclusively devoted to the feelings of envy. Different people have tried to explain jealousy in evolutionary terms (usually by referencing behaviors that seem more like a lion than a human – the man jealous because his female mate is pregnant by another male, thus thwarting his ability to pass on his own genes, the woman jealous because a straying man might not assist in the raising of the child, etc., etc.). Most of these explanations give rise to intriguing parlor conversations, but none of them has ever been useful for cloning a gene. Neither do they explain other aspects of jealousy that can only be directly associated with fecundity if you have a good imagination.

This lack of data might seem at first a bit puzzling, mostly because invidious feelings seem, as Dante eloquently points out, so obviously present. And historically troubling. Our capacity for envy has led to great wars and has changed the course of history. Reminiscent of our discussion on human avarice, we are forced to deal with something that appears to be a part of human nature yet may possess no direct biological correlates worth writing home about.

One comes to two conclusions regarding this lack of data on the biological basis of human envy: (1) our technology isn't good enough to detect the emotion system devoted exclusively to it, and (2) there is no such thing as an *exclusive* emotion system of jealousy. Perhaps envy is simply a human attempt at organizing subjective feelings that have no biological correlates at all. If the first conclusion is true, then we are going to have a very short chapter indeed. If the second conclusion is true, then perhaps the experience of jealousy – which appears quite real – actually has underlying associations that at first blush may not seem particularly obvious. If that's the case, then we may need to lift up a few rocks in envy's psychological landscape to see if anything biological comes scurrying into the open. If we find something, we might clearly see our next great lesson in attempting to understand human behavior: the presence of conscious feelings sometimes "obscures" our ability to understand emotions in a scientific fashion. That, I believe, is exactly what we will find when we take jealousy out from underneath the stones of human behavior and look at it under the biologist's microscope.

What we are going to study

Psychologists tell us that human envy, the unrequited feelings of discontent aroused by a desire for the possessions or qualities of another, can be easily associated with at least four sets of behaviors. Three of them we have already discussed – the behaviors surrounding sexual desire mentioned in Chapter Two, the behaviors of fear observed in the explanations of greed, and the feelings of aggression, mentioned in the previous chapter. In this section, we are going to discuss the relationship between envy and the fourth behavior, which is a human reaction to thwarted attempts at survival. Psychiatrists usually call this reaction depression. Note that I used the words "reaction", for depression can be as much a response to envy as a component. Note also that I used the words "thwarted attempts at survival". Many biologists view clinical depression, the most common form of all mental illnesses, in evolutionary terms. The very mechanisms that help us when we face immediate danger can actually hurt us if the threat feeling is long-lasting, and does not go away for a long period of time. And what could be more damaging than unrequited chronic longings? As we'll see in a moment, identifying a link between envy and depression is actually not far-fetched at all. It will also illustrate rather nicely how powerful that "obscuring" principle we just talked about really is.

Once the link between envy and depressive illness is established, we will be free to talk openly about the nature of clinical depression, its warning signs and psychiatric categorizations. Then it will be on to the biology. We will discuss the areas in the brain that may be responsible for the phenomenon, as well as the accompanying biochemicals and genes. We will conclude with some further reflections about what envy and depression tell us about minds and brains and consciousness, returning at the end to the detouring characteristics conscious feelings can have on the study of human emotions and human behaviors.

The relationship between envy and depression may best be observed intuitively, by considering a modern-day example. The subject of this example is infertility, a phenomenon that in Dante's time was often considered to be a curse from God. Susan, the twentieth century woman who experienced it, would probably agree.

The beginnings of an association

For as long as she could remember, Susan felt as though she had been adopted. In temperament, she was very different to most of her relatives; several were in jail for violent offenses, a few were alcoholic, two were in treatment for clinical depression. Susan was envious of the more "normal" families of her elementary school classmates. She became determined that when she grew up, *her* family would be very different. Susan would create a stable, sunny home environment, far from the troubled house in which she was raised.

She made a good start. Susan went to college, married an accountant and, after several years, started planning a family. But things slowly began to unravel as the couple realized they were infertile. After the first year of trying unsuccessfully to conceive, they went to see the doctor, who told them that nothing appeared to be wrong. Still she couldn't conceive, and while fertility clinics were available the couple couldn't afford them. Susan's frustration grew, and soon she was eyeing jealously the families she saw around her, the ones with kids, laughing in the park, playing in a sandbox, crying in a grocery store. She became envious of their happiness, their fertility, their seemingly stable *family* life, and soon found it hard to look at babies. Susan began withdrawing from her friends who had children. She even began to withdraw from her husband, becoming alternately sullen and angry towards him. Then she became angry with herself, at life, at God, knowing that she was starting to lose her grip on her dream of a stable household. The envy she had felt in elementary school came roaring back to her with a vengeance, and it wasn't long before Susan began to contemplate suicide. Her jealousy of her fertile friends was complete, her anger at her husband total, and Susan became locked in the grips of a potentially life-threatening depression.

The link between envy and depression

There are several aspects of Susan's life that we will be discussing as we explore the psychology of envy and the biology of depression. From a psychological point of view, a critical component to Susan's depression was her feelings of injustice – especially when she saw her friends with children – and her utter powerlessness to change her own situation. The perception of inequity and its role in depression has actually been studied in a formal fashion. So has the role of feelings of powerlessness. In one study, one hundred and seventy-five males and two hundred and fifty-two female college students were asked to journal their feelings of envy by selecting a particular person or persons of whom they were jealous. The subjects were then asked to rate how unfair they felt their situation to be (in both the subjective and objective sense), what the target's advantage was, how inferior the advantage made them feel, and either how hostile or how depressed the target's advantage made them feel. Using some complex mathematical analysis (and a few important controls) the psychologists came to the following conclusion.

> The envied person's advantage that made them (the subjects) feel inferior did not predict hostile feelings, but was strongly predictive of depressive feelings ... Envy may need to be understood as resulting in part from a subjective, yet, robust sense of injustice.
>
> *(Smith* et al., *1994)*

As a molecular biologist, I must necessarily scratch my head at some of the methodologies, definitions and conclusions that are derived from studies like these. I also understand that the subjective nature of certain types of human behavior (like envy) can be very difficult to study in an objective, quantifiable fashion. Indeed, taking up this subjectivity is the point of this chapter, illustrating that the presence of subjective conscious feelings sometimes clouds our ability to understand emotions in a scientific fashion.

Yet, there are many studies linking subjective feelings such as envy to
honest-to-God biological processes like depression, and so the association
between the two must be addressed, even if the scientific borders are ill-
defined. Another study looked at suicides over the human life cycle, examining risk factors, assessment and treatment of suicidal patients, many of
whom displayed dark feelings of envy, jealousy and resentment. The conclusion of this work was the same as that of the study mentioned previously.

> It is not rage and depression alone that must regularly be addressed in
> therapy, but underlying dispositions to envy, rivalry and jealousy . . .

It appears that feelings of powerlessness to change an unhappy situation are very much a part of human depression. This hints at survival issues.
Indeed, as we'll see in a minute, there are researchers who believe that
depression is simply a part of our "fight-or-flight" emotion system gone
out of control. And if that's the case, then envy may simply be a subjective
reaction to the feelings of helplessness, and depression the deregulated
emotion system that undergirds it. Getting at the underlying feelings of
jealousy really means addressing this powerlessness, which, in the view of
the study just mentioned, is part of the road to healing.

Powerlessness certainly seemed to be the case with Susan. As her inability to conceive was emotionally tied to her well-being, even to her sense of
"pulling herself up by her bootstraps", not being able to fulfill those goals
left her emotionally abandoned. Seeing other couples with children ignit-
ed the incendiary jealous feelings, acting as an environmental stressor,
reminding her that there was no way out. Since depression already ran in
her family, it is quite possible she was predisposed to it, a notion we will
address when we discuss the genetics of depression. Indeed, Susan's story
does not have a happy ending. As her depression mounted, she became an
insomniac, and her doctor prescribed sleeping pills to get her through the
night. One morning she decided to overdose on those pills. She was dead
by the time her husband came home from work.

However one wishes to make the link, the association between depression and envy is evident, palpable, even lethal. That's why we can start out
talking about jealousy and so easily end up describing suicides.

What depression looks like

The events that led up to Susan's suicide are in many ways typical of depressive illnesses. Over the years, there have been attempts to categorize the types of depressions human beings experience based on clinical histories. There are long-term depressions, for example. These are usually classified as "low" feelings that have lasted for more than two years. People who suffer from this kind of depression often have never known a time in their life when they weren't depressed. Psychiatrists call this type of depression dysthymia. There is also single-episode depression. This is the kind of illness that strikes a person literally out of the blue. One moment the person is feeling fine and the next moment everything turns black. What's curious about this form is that the individual has no history of depression, and, once treated, may never suffer from it again. The most common form of depressive illness is the so-called recurrent depression. People who suffer from this form often have sharp, severe episodes of depression. The bout can be triggered by an environmental stress, such as the inability to have a child, although the depression can be incited from sources that do not appear obvious.

There are other forms of the illness as well, some of which don't even seem like depression. There is a kind of depression that also has a "manic" side to the behavior. This manic side is characterized by excessive energy, feelings of elation, incoherence, and disorganization. People who possess this manic characteristic often cycle back into a deep depression, and then alternate between the highs and the lows. We used to call such disorders manic depression, though the more common term these days is bipolar disorder. Both depressive and bipolar illnesses are classified by psychiatrists as mood disorders.

How can you tell depression from normal "blue" feelings?

At one level, bipolar disorder is easy to study because the symptoms are so dramatic, and so cyclical. But the vast majority of people who suffer from clinical depression do not present bipolar symptoms, but characteristics of recurrent depression. And that makes things a little more difficult, because humans can go through "down" periods in their lives and not be in the grips of a disease state. Susan might certainly have gone through the common ups and downs of life, but it wasn't until the infertility issues crept up in her life that she exhibited clinical disease. And that brings out an important point. Depression should not be mistaken for everyday feelings of sadness, nor should it be confused with the intense feelings of grief associated with the loss of a loved one. Such emotional lows are the normal reactions to the tensions of every day. Time goes by, black clouds give way to sunshine and life goes on.

People who suffer from depression often do not feel better for many months, years – in some cases even decades. Susan did not commit suicide for almost three years after she tried to get pregnant, for example. In an effort to distinguish between normal feelings of loss and depressive illness, psychiatrists have created a ten-point checklist. In general, anyone who suffers from five or more of these symptoms nearly every day, all day and for more than a fortnight, probably has an illness that requires treatment. The important point is evaluating the duration of the symptoms, and asking whether they interfere with a person's ability to function. Here is the ten-point checklist.

1. Loss of energy, fatigue, feeling "slowed down" or "dragged down".
2. Feelings of emptiness, persistent, low feelings of anxiety.
3. Loss of interest in pleasurable activities (a loss of interest in sex is one bellwether of depressive illness, for example).
4. Disturbances in sleep patterns (insomnia or oversleeping, even early morning wakening, for example).
5. Appetite and weight changes. These can either be gains or losses.
6. Recurring thoughts of death or suicide.
7. Feelings of worthlessness, guilt, feelings of being trapped.
8. Memory lapses, loss of concentration, inability to make decisions.
9. Pessimistic, hopeless feelings. These are most often the feelings that we associate with the normal ups and downs of life. These are feeling of intense despair, however.
10. Physical symptoms of disability, pain not caused by physical disease (headaches and stomach aches, for example).

A cursory examination of this multi-point checklist seems to point towards two overall kinds of symptoms in human depression. One type seems to be composed of subjective feelings. There are guilty feelings, or there are feelings of worthlessness, the sense of emptiness and anxiety, the hopeless feelings of being trapped. Because of their subjectivity, many people think that depression is a personal "fault" rather than a disease. Such subjective notions as fault seem about as amenable to scientific inquiry as the concept of envy. And indeed, if all there was to our understanding of human depression were subjective insights, progress in uncovering the biology of the disorder might have been a lot slower.

There is, however, another set of symptoms in our clinical depression list that don't seem to be subjective at all. There are changes in sleep patterns, alterations in weight, the inability to remember things, the loss of energy, changes in libido and so on. These symptoms are different to the other more subjective components in the list in that they can actually be measured. As you recall, neurobiologists have gotten a handle on where in the brain such behaviors are mediated. The hypothalamus, for example, is deeply involved in weight regulation and our ability to use energy resources, in sleep regulation and our ability to feel energized, in our sexual feelings, even our ability to process and respond to fear. Since depression is often accompanied by changes in this suite of behaviors, scientists arrived at a powerful working hypothesis: perhaps depression is the result of a biological problem. And maybe part of this problem is centered in the hypothalamus.

In the next section, we are going to discuss the biological underpinnings of human depression, including a discussion of the role of the hypothalamus in major depression. There are three overall theories currently used to explain the biology of depression, each centering around a different brain system. It is possible, however, that all three theories describe separable aspects of a single phenomenon. To describe what I mean, I would like to leave the world of twenty-first century neurochemistry and return to Dante's Florence, describing an activity that serves as a useful metaphor for the theoretical descriptions of depression. The metaphor comes from a stanza in Cantos XIII, which describes the apparel of the inmates trapped in the dark world of the Second Cornice.

An introduction via textiles

Florence, as we have discussed previously, was known in Dante's time for its fine clothing manufacture. Woolen cloth was the primary source of clothing and entire guilds were built around the manufacture and distribution of textiles and apparel. Because of this emphasis on fine fabric, the implications of the following reference would not be lost to a Florentine knowledgeable about the business:

> For when I drew near and could see the whole
> Penance imposed upon those praying people,
> My eyes milked a great anguish from my soul
> Their cloaks were made of haircloth, coarse and stiff.
> Each soul supported another with his shoulder,
> And all leaned for support against the cliff.

The word that would stand out to a man from Florence in the thirteenth or fourteenth century would be *haircloth*. Especially a big cloak of the stuff and against bare skin. Haircloth is an extremely heavy, coarse fabric made from goat hair. Its dense weave makes it ideally suited for warmth, and haircloth is still worn by the rural inhabitants of Northern Italy in the winter months. In Dante's time, haircloth was worn against the skin either for penance or to inculcate a form of spiritual discipline, a kind of holy torture, if you will. It was used because of its insufferable itchiness and its ability to create surface wounds. The cloth was so coarse that it usually tore the flesh open when rubbed against the skin. The end result was that in Medieval times people who wore haircloth against their skin often had running sores. The dense nature of the cloth resulted in its becoming a home to all manner of insects, especially fleas. These would of course feed on the open wounds, and, since cleanliness was not often a feature of the medieval toilette, the haircloth became kind of an arthropodian Noah's ark. Bacterial infections, including gangrene, were common results. Wearing haircloth was thus a form of torture, and, for those poor blind envious souls of the Second Cornice, a source of their punishment.

The most interesting aspect of this ghoulish cloth is the cumulative nature of interlocking events. In order to get the gangrene, several things had to be working together. A person might be forced to wear haircloth, for example, but no infection would be present until the skin became abraded. Once an open sore was available, however, it was vulnerable to any germ-laden piece of dirt or spittle that might fall into it, creating the infections. Germ-carrying insects could be a source of trouble also, but the diseases mediated by them would not start until the creatures themselves set up residence. The haircloth thus served as a kind of nasty crossroads where many sources of infection might converge. The end result was the same – infected human tissue – but the resulting pathology could be derived from several, loosely related sources.

I would like us to remember this crossroads role of the haircloth as we begin to explore the biological underpinnings of human depression. As we shall see in a moment, human depression may be associated with changes in as many as three separate brain systems, each by themselves capable of causing disease. Recent evidence suggests that, in some cases, these systems may be interlocking, working together to create the pathology. In other words, there may be the neurological equivalent of haircloth wrapped around our brains which, if rubbed in the right way, could cause a cognitive "infection", a depression. And there may be several ways to start the infection process.

This impression of a biochemical cause of depression has come after excruciatingly difficult research efforts executed over many years by literally hundreds of people. The first clue that a physiological process might be involved came from subjective genetic impressions. You may have noticed from the description of Susan's background that depression seemed to run in her family. Many clinicians have found a similar pattern for their depressed patients. Indeed, when a psychiatrist first interviews a person who appears to be suffering from depression, one of the first questions they ask is, "Is any other member of your family currently being treated for depression?"

We will begin our exploration of this haircloth model of depression by exploring the answer to the psychiatrist's question: looking at the search for the genes of human depression. We shall find that this direct effort has mostly met with failure. But that does not mean that a biological explanation for depression is not within our grasp. Indeed, there is more reason than ever to think that we may find gene sequences for depressive illnesses. After discussing the search for genes, we will explore some of the data that give rise to this optimism, and see exactly how a haircloth wrapped around a neuron may help explain why Susan overdosed on her sedatives.

The hunt for a gene

The fact of repeated disease within family members makes everybody think of isolating the genes. Certain researchers have devoted their entire careers in an attempt to find one for human depression. But sensing that a gene may be involved and isolating the offending sequence are two very different things; finding any gene for a specific behavior is one of the most difficult feats in all of research biology. The technique that researchers use to find such sequences is worth reviewing here, illuminating this difficulty, and showing why, after all these years, we still don't have a gene for the disorder.

One of the most common techniques researchers use to isolate behavior genes involves establishing something called a LOD score. A LOD score is actually a measure of genetic association, or, in the parlance of the researcher, genetic linkage. A complex ratio complete with important formulae, the LOD score in essence describes the probability that a given area of a chromosome will be linked to a given behavior trait. The technique can be divided into three general steps.

Step 1. If a pattern of recurrence of a disease is sufficiently regular within successive siblings and generations, a particular genetic mutation (or perhaps series of mutations) may be inferred. While that may sound conceptually simple, it is easy only if the trait being sought has a strict definition. For diseases like cystic fibrosis, which has a rigidly defined clinical profile, discovering who has the disease and who does not is straightforward. But in the case of human depression, obtaining a strict profile is not necessarily an easy thing to do. You may remember from the beginning of this chapter that there are several categories of depressive illness. Which set of criteria is the *right* one to study?

Step 2. This involves examining the genetic material of affected individuals to see if there is any abnormality. The abnormality may be as conspicuous as a giant chunk of chromosome missing in the patient who exhibits the disorder. The abnormality may also be quite subtle. Perhaps only a single nucleotide difference separates a normal person from one who has the disease. Maybe a gene that is unrelated to the disorder but whose physical placement is close to the gene for the disorder is abnormal. Whatever the mutation turns out to be, the second step is to find an associated physical anomaly.

Step 3. The third step is now to relate steps one and two to each other. That is, you must correlate the presence of the chromosomal abnormality with the presence of the disorder criteria in a family of patients. The more patients you can test, the better off you'll be. This often means investigating entire family trees, testing as many living relatives as possible, trying to determine who has the disorder and who does not. Next you have to make the association between the biochemical abnormality and the disease. You must demonstrate that this correlation is not related to chance; that is you must show that every time someone gets the genetic abnormality, they also get the disease profile. This can be very difficult to do, especially if the family trees are not very reliable. It is one of the reasons why researchers often use specific people groups who have kept to themselves reproductively, and have also kept good records of their families (the Amish community in Pennsylvania is one example of a group that has been studied extensively).

This associative comparison is done formally and the mathematics involved establishes the LOD score, that is the likelihood that there is a linkage between the disorder and the chromosomal abnormality being examined. The higher the LOD score the better; for example, a LOD score of 1 is not considered to be very good, whereas a LOD score of 8 is considered to be quite good. Most people use a LOD score of 3 as the cut-off point.

So how is the search progressing?

The LOD score protocol is not the only technique used to isolate specific genes in populations of people. Indeed, there are now several statistical/ evaluative methods geneticists use to try to clone these important sequences – and most of them have been tried on this problem of getting at depression genes.

Unfortunately, the search has not gone well, regardless of what method is employed. You may recall from the last chapter how scientists use twin studies in an attempt to define the initial boundaries in the border dispute between genetic and environmental territories. For bipolar disorder, it has been shown that the prevalence of the illness in both members of a pair is greatly elevated in identical (monozygotic) when compared to fraternal (dizygotic) twins. A similar, though much less dramatic elevation, is seen when one examines particular types of major depression.

That's not a promising start and, perhaps predictably, there have been many dead-ends. A study of Amish people and bipolar disorder initially localized a sequence of DNA on chromosome 11. But as more work was done (and more Amish relatives studied) the data became awash with background noise, with the LOD score for chromosome 11 eventually dropping to chance levels. Other suspects have come to the fore, notably the X chromosome, chromosome 18 and even chromosome 21. Most of these studies have either not been replicated or await confirmation. At this time, the hunt for the gene for this mood disorder, is, well, depressing.

Why is it so difficult?

The reasons for this lack of success are manifold, and are very instructive about how to view genes and behavior. As we stated in the introduction, the motivation for finding a simple explanation is powerful, even for something as complex as human comportment. The pitfalls are worth going over, and, before we leave the subject of genes and behaviors, I would like to describe why the pitfalls exist. Here are just four of the many reasons why this research is so tricky.

Environment. This is the obvious nature/nurture debate now applied at the most intimate level possible. The recurrence of a particular psychiatric disorder within a family does not always mean that there is a purely – if any – genetic reason for its existence. As you recall from the beginning of the chapter, there are depressions that seem to come right out of the blue and never show up again after successful treatment. How might that occur? There may be a certain roll of the genetic dice which may induce the disorder only if the right environmental stresses are present. So, if those stresses are not present the disorder may never show its clinical stripes, *even if the person has a mutation capable of causing the disease*. So you might get the disease once and only once, because of some unfortunate combination of environmental circumstances working with this more hidden mutation. In terms of genetic analysis, this kind of thing can really screw up the numbers.

Multiple origins. Another problem is that any one of several genes may all cause the same disorder. Thus we might be hot on the trail of one kind of mutation, linking a certain gene with a certain disorder, and then someone comes along who has all the clinical manifestations of the disorder we're studying and, when we examine the suspect gene, we find nothing wrong with the DNA. Conversely a person might have the suspect gene but not manifest the disorder at all. Why? One explanation is that a number of mutated genes could all cause the disease one is looking for, and different people, even within the same family, might be carriers of different DNA anomalies. This misidentification of disease with genes is exactly the kind of thing that occurred in the Amish study with bipolar disorder. A particular region on chromosome 11 was studied in more members of the family tree, and it was found that there were those who had specific mutations, but not the disease. At the same time, there were those members who displayed the disease, but did not have the specific mutations. Either way, the numbers become murky very quickly. One explanation for this phenomenon is the idea of multiple genetic origins.

Polygenic origins. Even if a disorder has a specific genetic basis, there may be two or more mutated genes that must be present before the disease profile is observed. This gives you the same problem encountered as above. A scientist might be examining a single mutation in a family, thinking that they have isolated a target gene, never knowing that a necessary second gene was segregating along with it, perhaps in tight association, but nonetheless different. For example, we may eventually find that a mutation exists in our suspect gene in Patient A, and yet find that Patient A has absolutely no clinical manifestation of disease – the explanation being that Patient A did not get that necessary second gene from his or her parents. This can lead to all kinds of false trails, making the need for accurate family trees with lots and lots of siblings very important indeed.

Classification. This anomaly we have already mentioned. The ability to create an accurate profile of a disease, a profile that makes sense to a test tube, may be a problem. Words like bipolar disorder and depression may in the end just be catch-all phrases for many groups of disorders. Some of these disorders may be related and some not. Since one has to create an exacting profile at the very outset of the isolation attempt, an early mistake dooms the research effort quickly. One of the great tasks of molecular biology is to create categories of pathologies that are amenable to scientific inquiry.

At this point in our discussion, you may be wondering how anybody can find strong biochemical correlates to behaviors such as depression. As you can see from the page numbers, you are only about half way through this chapter and, in fact, there is a great deal of hope in understanding the biology of depressive illness. It is just that using techniques such as LOD scores and other types of analysis have not borne fruit quickly.

Other research using different techniques *has* borne a great deal of fruit in our hunt for the genes of depression, and it is to this that I would like to turn our attention. The data come from looking at the roles played by specific regions of the brain in our various moods, and then examining the resident hormones and chromosomes.

To understand how various regions in the brain work to create human depression, I would like to use a military analogy, focusing on the hierarchical chain of command seen in the armies of thirteenth century Italy. This particular example starts with a comment from a woman Dante met early in his journey in the Second Cornice. The woman was Sapia of Siena, the wife of a prominent nobleman, and paternal aunt to her noble nephew, a man named Salvani.

The professor lectures about the battle of Montaperti

"She watched her nephew Salvani being beheaded after he lost the battle," the professor explained, beginning his description of one of the most important conflicts of thirteenth century Florence. He patted his neck, warming to the subject of medieval army organization. "She *loved* the sight of his death, which occurred at the behest of the victors, ever demon-strating the awesome power of the medieval chain of command. Sapia watched the whole thing from a palace window. This is what she suppos-edly said when her nephew was dead:

> Now God, do what you will with me, and do me any harm you can, for after this I shall live happily and die content.

The professor explained that the reason for such perverse peace had to do with political rivalry in the middle of medieval Italy. History has dubbed the conflict Sapia witnessed the Battle of Montaperti, in which two hope-lessly envious Italian cities, Siena and Florence, were pitted against each other in mortal combat. "Her nephew was Provenzano Salvani, one of the military chiefs of Florence," he continued. "As her name suggests, she was from Siena, the enemy city on the other side of the conflict. She hated Salvani and she was scared of him, because the Florentine army of which he was a part was much larger than her native Siena's."

It is questionable whether you could actually call the enemy army Florentine. The professor began to explain the extraordinary components and chains of command that existed within this thirteenth century mili-tary force. "It was really a cooperative adventure possessed of many kinds of people," he said. "The Florentine army had professional soldiers, citizen soldiers and mercenaries from God-knows-where hired for the specific occa-sion." That didn't mean the thing was disorganized, he cautioned. "There were cavalry, archers, crossbowmen, sappers, and infantry, each with a leader and cadres of messengers. And there were companies led by com-manders, all of whom were in touch with the messengers and sub-lieu-tenants. So it was not too disorganized. But the interesting thing was that the commander might not be from the city doing the fighting. He might be hired from a neighboring region, employed for the battle simply because he had experience fighting battles."

That experience meant that the commander was used to a chain of com-mand, even if he supervised a polyglot of troops like the Florentines. The professor continued, "If he wanted to move a company of foot solders to such-and-such a hill, he would use an apparatus not all that different from that used today. The commander would issue an order to a sub-lieutenant, who might ride out to the captain of the company, and the men would move." The commander might observe the effects of his decisions from some high point, or he might rely on feedback from various messengers. "Based on what he saw," the professor explained, "he might issue an order to charge, to retreat, to move, or whatever." The importance of the hierarchy was preserved, and to great effect for Siena in the Battle of Montaperti. Against overwhelming odds, the Sienese soldiers slaughtered ten thousand Florentines and took another twenty thousand prisoner. Part of the reason for the lopsided victory was the polyglot content of the Florentine army.

"But even after the battle, the issuance of orders was a big deal. And in the time of Dante, such commands could be sickening." This was certainly true in the case of Provenzano Salvani, the nephew-leader of the Florentine opposition. He survived the battle and was taken prisoner. "The command came forth that he be executed immediately, and, in obedience to those orders, he was beheaded right there on the battlefield. His aunt, this Sapia woman, then uttered her famous sentence. Such is the consequence of the medieval hierarchy." The professor tapped his neck one more time.

The relevance of military hierarchies to depression

The defeat of the Florentines at Montaperti did not mean the end of the city of Florence. She was able to field other armies, with thankfully clear-er chains of command, and rose to become the most influential city in Tuscany. But lest we lose sight of the subject of this book, I am bringing up the Battle of Montaperti in a chapter devoted to human depression for a specific reason. In many ways, the individual groups of neurons respon-sible for depression are arranged like the chain of command in that polyglot thirteenth century Florentine army. I would like to use the struc-ture of this chain of command as an analogy in continuing our discussion of depressive illnesses. To see what I mean, consider how one brain system involved in depression interacts with specific regions of the body to create a down-in-the-dumps effect.

This brain system starts with a single general, just like the Florentine army at the Battle of Montaperti, in charge of an extraordinarily diverse, even far-flung group of neural tissues. The general is a region of the brain very used to giving orders, a region so familiar to you now that you might be tempted to think it the only important group of neurons in the whole brain. I am talking, of course, about the hypothalamus. Like the top offi-cer, this region receives input from the environment via other regions of the brain, and if it senses stress the hypothalamus will issue a series of battle commands.

As we discussed in Chapter Four, on greed, these commands come in the form of secreted hormones. One of the most powerful of these biochemicals we mentioned is called corticotrophin-releasing factor (or CRF). As you recall, when the hypothalamus senses stress it secretes CRF. This CRF acts like some sub-general receiving instructions from the overall commander of the Florentine army. CRF doesn't travel very far, but delivers its message to another general in the chain of command, the pituitary gland.

Once CRF arrives at the pituitary, it tells the gland to secrete yet another powerful hormone. You might remember that this hormone goes by the name adrenocorticotropic hormone, or ACTH. This ACTH acts like a traveling messenger boy in the Florentine army, one who carries the commander's set of instructions to the relevant part of the fighting force. ACTH does not travel to some distant hill, however. As you recall, it is secreted into the bloodstream, where it ventures far away from the brain, arriving at a series of tissues that lie on top of our kidneys. These tissues are called the adrenal glands.

What in the world would a brain chemical have to do with tissues perched on our kidneys? As we have already discussed, adrenal glands actually have a large number of functions related to human behavior, even though they are quite distant from the brain. One of their functions is to respond to the brain by making important chemicals, most of which are dumped back into our bloodstream. When ACTH arrives at our adrenal glands, that is exactly what happens. The adrenals secrete one last hormone, a very powerful molecule we did not mention in Chapter Four. It is called cortisol.

What does cortisol do? It performs many missions in our bodies, most of which relate to the flight-or-fight responses we have been discussing. Cortisol travels to the brain via the bloodstream, a lot like a messenger sent back from the front to inform the medieval commander about the situation in the battlefield. Cortisol suppresses our desires for food and sex, which are energy-wasting and irrelevant behaviors in a stressful situation. At the same time, cortisol increases our overall alertness. It also elevates the delivery of fuel to our muscles, a needy supply for exercise should we require additional energy. Taken together, cortisol and other hormones prepare the body for attacking or retreating, all in an effort to preserve the life of the person feeling the stress.

The last thing these hormones do is as important as the preparatory functions just described. They perform an inhibitory service for the various generals in the chain of command. They inform the pituitary and the hypothalamus, for example, that the body is prepared and executing functions, and that there is no need to send out for additional reinforcements. This prevents the manufacture of the various signaling hormones from becoming excessive. What occurs then is a feedback system reminiscent of many of the loops we have discussed before. The brain sends out chemicals to perform certain functions, and these functions stimulate other chemicals to come back to the brain, telling it to relax.

Taken together, we have the cooperation of a polyglot of tissues in our body, all capable of reacting to stress, with hopefully better final results than the army of Florence assembled at Montaperti. The brain feels the stress, relays to the hypothalamus, then to the pituitary, finally to the adrenals. This interesting hierarchy of organization is often called the HPA axis, taken from the first letters of each of the tissues involved (see Figure 21 for a detailed explanation).

The relevance of the HPA to depression

So far in our discussion we have only addressed the body's preparation for stressful situations. What does this preparation have to do with human depression? The answer appears to be both "a great deal" and "we don't know", the same familiar comments we've encountered from other subjects addressed in this book.

The "a great deal" comment mostly comes from associative data, complete with all the pitfalls we mentioned in our discussion of testosterone and aggressive behavior. It has been known for more than thirty years, for example, that depressed people tend to have excess levels of cortisol in their bodies. This is true whether one examines their urine, their blood, or even the fluid that bathes their brains and spinal cords (called the cerebrospinal fluid). When one examines the tissues in the HPA axis, one sees similar hyperactivity. Depressed people tend to have enlarged adrenal and pituitary glands, indicative of overwork. These overworked adrenal glands tend to secrete more cortisol than those of individuals not experiencing depression.

The same kinds of hyperactivity appear to exist in the "command and control" center of the hypothalamus as well. When one examines the neurons responsible for creating hypothalamic CRF (which is, as you recall, the inaugurating hormone for the stress response), one sees that they are working overtime. In depressed individuals, the gene responsible for the creation of CRF is hyperactivated in these neurons. There tend to be more of them (neurons) in depressed individuals. Depressed people also tend to have more CRF in their cerebrospinal fluid compared to unaffected individuals. The list seems to go on and on regarding the association between various hormones of the HPA axis and human depression.

Exactly what is the HPA axis?

The HPA axis stands for a relationship between the hypothalamus, pituitary, and adrenal glands. They work together to prepare the body in the presence of a perceived threat. Here's how it works.

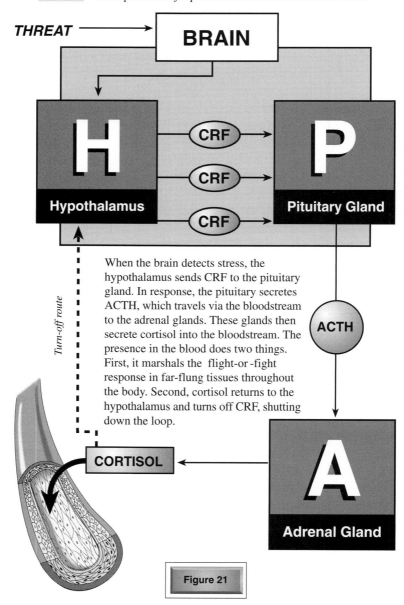

When the brain detects stress, the hypothalamus sends CRF to the pituitary gland. In response, the pituitary secretes ACTH, which travels via the bloodstream to the adrenal glands. These glands then secrete cortisol into the bloodstream. The presence in the blood does two things. First, it marshals the flight-or-fight response in far-flung tissues throughout the body. Second, cortisol returns to the hypothalamus and turns off CRF, shutting down the loop.

Figure 21

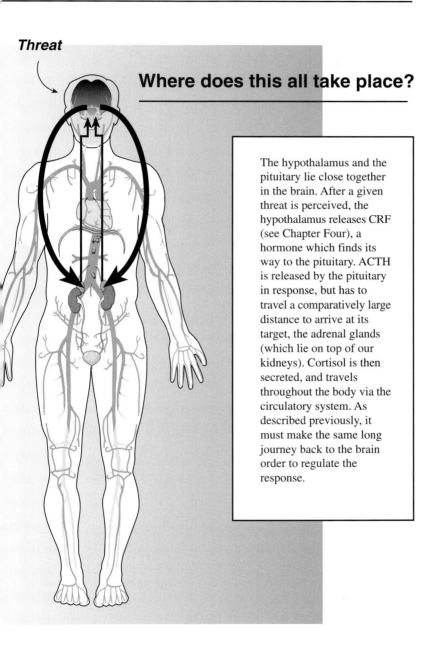

Threat

Where does this all take place?

The hypothalamus and the pituitary lie close together in the brain. After a given threat is perceived, the hypothalamus releases CRF (see Chapter Four), a hormone which finds its way to the pituitary. ACTH is released by the pituitary in response, but has to travel a comparatively large distance to arrive at its target, the adrenal glands (which lie on top of our kidneys). Cortisol is then secreted, and travels throughout the body via the circulatory system. As described previously, it must make the same long journey back to the brain order to regulate the response.

But what about depression?

All the tissues discussed above are related to the stress an individual feels when encountering a threat. But this is not a chapter about fear, which means we have not yet addressed the question we asked in the previous section. What about depression? How does the HPA relate to people who want to commit suicide? This is where the comment "we don't really know" comes into play regarding the brain systems we just mentioned. The data presented here are primarily associative, even though, in some cases, thousands of experiments buttress the claims. As we shall see in a few minutes, multiple factors may be involved in the creation of human depression. And when those factors are teased out from the behavioral data, we shall probably have many types of depression, each with their own physiological explanation.

One popular hypothesis that involves the flight-or-fight response comes from an interesting experiment done with rats. When large amounts of CRF are applied directly to the brains of laboratory animals, they exhibit symptoms very characteristic of depression. They eat less and sleep less and have greatly reduced sexual activity. They are increasingly restless in familiar surroundings and tend to retreat in unfamiliar environments. It is possible that certain types of depression occur because the normal regulatory pathways of the HPA have been altered. It's almost as if the people suffering from depression can't shut off the "Oh-my-God-here-comes-the-saber-tooth-tiger" reaction, even though the tiger left the vicinity months ago. CRF continued to be pumped to the pituitary gland, resulting in the adrenals making more and more cortisol. Since cortisol alters sleep, appetite, and sex behaviors – all the symptoms depressed people feel – one might predict from the hypothesis that depressed people would (1) have elevated levels of CRF and (2) possess elevated levels of cortisol. As we have seen, that is exactly what is observed. With no shut-off system available, the person slips into a depressive illness.

Admittedly, the fight-or-flight hypothesis of human depression makes an attractive explanation. But it must remain in the realm of hypothesis, not only because much of the data is associative, but also because there are many pieces of the puzzle that do not yet fit. Chief amongst these misfit observations is that some kinds of depression appear to occur – and can be successfully treated – by altering molecules that don't seem to be related to flight-or-fight at all. There are as many as three separate brain systems involved in human depression, only one of which is the HPA axis. We are now going to consider another system which centers around a class of biochemicals we have mentioned before. The class is called neurotransmitters, and the specific neurotransmitter we shall focus on is called norepinephrine.

What words like noradrenergic mean

As you recall, neurotransmitters are made to help neurons communicate with each other. Neurons form linkages together, creating what are in effect complex highways coursing through the brain. Many neural highways use only one class of these neurotransmitters as their primary communication agent. The neural highway that uses norepinephrine is termed the noradrenergic system.

The trick to understanding neural communication throughout highways like the noradrenergic system is to know how neurotransmitters work. You might recall from Chapter Four that we discussed their role in some detail. We talked about the fact that neurons are physically separated by a space termed a synapse. You might further recall that in order for one neuron to communicate with another, this space must somehow be bridged. This bridging occurs with the use of neurotransmitters. One neuron, when stimulated, will secrete neurotransmitters into the space. The tiny biochemicals travel across the synapse and land on receptors embedded on the surface of the receiving neuron. Such binding can elicit changes within the receiving neuron, changes that might stimulate it (or inhibit it). In either event, information is transferred. We used the analogy of two medieval port cities separated by a watery Adriatic Sea, Ancona and Zadar, to illustrate how this transfer occurs.

But how do we know that norepinephrine is involved in depression? There is a great deal of experimental support linking the neurotransmitter to the mood disorder. The role of the neurotransmitter in depression was actually hypothesized some time ago. In the 1960s, Dr. Joseph Schildkraut of Harvard proposed that humans experience depression when norepinephrine levels are depleted in the neurons that normally use it for communication.

Exactly how might those levels become depleted? Understanding how neurotransmitters work, and what may happen to them in the midst of a depressive illness, is a fairly complex task. I will attempt to cut through some of the complication with a familiar analogy, extending the idea of medieval seaports discussed in Chapter Four. Specifically we will focus on the shipping activity of one such city, busy today as in Dante's time, the seaport of Ancona, and see how its activity helps us better understand norepinephrine and depression.

A visit to Ancona

I have had the distinct misfortune of visiting contemporary Ancona, a town on the Northeastern coast of Italy, and still the largest seaport on the Adriatic. My negative comment comes from the fact that it was ravaged seemingly beyond repair by allied bombing in World War II. And what the bombs did not destroy was finished off by a massive earthquake in the 1970s. Ancona has been only marginally reconstructed; with noisy trucks rumbling past dirty, collapsing buildings, the city is now used primarily only as a port and ferry terminal. That's not to say that Ancona is irredeemable. A beautiful and steep hill still overlooks the port, the bluff reminiscent of a bent elbow (hence the word Ancona, literally "elbow" in Greek). And there is the amazing Adriatic, still mysteriously gray and blue and fetching after all these years.

Unknown to bombings and earthquakes in Dante's time, Ancona would also have been a bustling seaport, its activities as recognizable to us in the twenty-first century as to a Florentine merchant in the thirteenth. There might be ships unloading their goods at one dock, and at another dock a ship loading its cargo, getting ready to cross the Adriatic. There might be a place where a damaged boat could be hauled ashore for repairs, or dismantled in a salvage operation, cannibalizing its parts for a future vessel. The boats traveled far and wide, and the port was used by the merchants of Florence for export and import purposes. In our analogy, we traced one such activity to a seaport on the other side of the Adriatic, Zadar, in present-day Croatia. One can still see the magnificent *Loggia de Mercanti* on the streets of Ancona, the business nerve center of commerce, built in medieval times, now caked with dirt and grime.

The amount of trade in a given seaport influenced its importance, of course, both from a prestige and a financial point of view. The number of usable docks a seaport might possess, its repair facilities, and the harbor's ability to remain unsilted and usable were all critically important in gauging a city's influence and power. It is no wonder that most of the greatest cities of the Middle Ages were connected to saltwater in some fashion. The effect was transferable. Boat traffic arriving in and out of a given port determined the economic health of the surrounding countryside in many important ways.

The relevance of Ancona to depression

It is this level of trafficking, the notion that activity influences strength, that I would like to capitalize upon as we seek to better understand molecular neurobiology, especially in the context of depressive illness. In many ways, the levels of neurotransmitter between neurons and their effects on human thought work like shipping activity between ancient seaports, and I bring up boat trafficking specifically to discuss norepinephrine's role in depression.

As already discussed, a great deal of evidence suggests that declining levels of norepinephrine between specific neurons are important in mood disorders like depression. What do I mean "between specific neurons"? As ever, there is a presynaptic neuron, the cell that secretes the neurotransmitter, and, across a saltwater expanse, a postsynaptic neuron set up to receive the neurotransmitter. As explained in Chapter Four, this reception occurs when a neurotransmitter from the presynaptic cell binds to a receptor on the postsynaptic cell, just as a medieval ship leaving Ancona might eventually tie up at a loading dock at Zadar.

So how does a cell lower the amount of norepinephrine in the synapse? And how does that create depression? The simplest way to alter the effects of norepinephrine is to lower the number of molecules being secreted into the synaptic space. If Ancona wanted to disrupt trade to Zadar, for example, it might allow only a few ships to leave its port. Another simple way to lower the amounts would be to let a lot of ships go, but then quickly recall them *back* to port in the middle of their journey. Believe it or not, presynaptic cells are capable of doing this. Many have what are termed "autoreceptors" on their surfaces. Once the cell releases norepinephrine, some of the neurotransmitters immediately turn around and sail back to the presynaptic origin, binding to these autoreceptors rather than making the full journey to the postsynaptic cell. This binding sends a signal to the presynaptic cell telling it that there is enough of the stuff around, and to quit making and/or secreting any more new neurotransmitter. In such fashion, not many neurotransmitters make the journey, which has the effect of lowering the amount of neurotransmitter available to a given postsynaptic cell.

Another way a presynaptic cell can lower the amounts of available neurotransmitter is to simply capture any postsynaptic-bound neurotransmitter ships and haul them inland, bringing them into the presynaptic cell's cytosol. Once inside, the neurotransmitter may be disassembled and its parts cannibalized for other functions, just like a repair dock in a

Depression means altering neural communication

Neurons normally communicate via biochemicals termed neurotransmitters. Depression occurs, in part, when this communication is disrupted. Shown below is a summary of how neurons normally use neurotransmitters to communicate. Illustrated on the next page are four ways in which this communication can be disrupted.

A The presynaptic neuron receives information in the form of electricity. As a result, the nerve "fires".

B In response to this firing, the vesicles containing neurotransmitters begin migrating to the edge of the nerve.

C These vesicles fuse with the neuron's outer membrane. As a result, the neurotransmitters are dumped into the synapse.

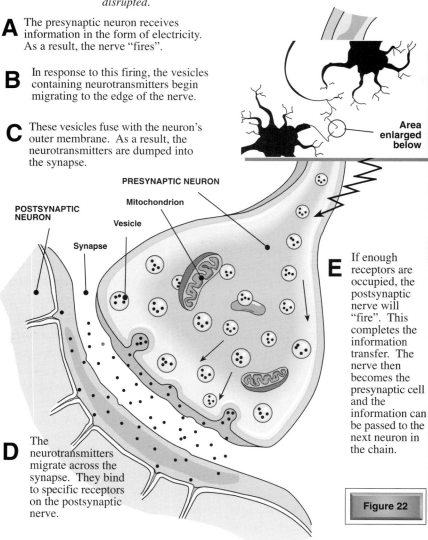

Area enlarged below

PRESYNAPTIC NEURON

Mitochondrion

POSTSYNAPTIC NEURON

Vesicle

Synapse

E If enough receptors are occupied, the postsynaptic nerve will "fire". This completes the information transfer. The nerve then becomes the presynaptic cell and the information can be passed to the next neuron in the chain.

D The neurotransmitters migrate across the synapse. They bind to specific receptors on the postsynaptic nerve.

Figure 22

WAYS OF DISRUPTING NEURAL COMMUNICATION

1 The presynaptic cell secretes a lowered amount of neurotransmitter into the synapse.

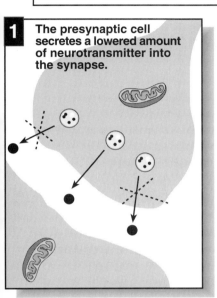

2 Secreted neurotransmitter binds to an autoreceptor on the presynaptic cell rather than on the postsynaptic receptor.

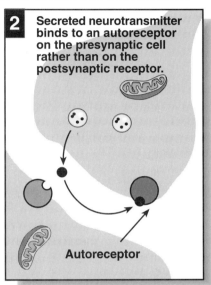

Autoreceptor

3 In a process known as re-uptake, the presynaptic cell internalizes its own neurotransmitters.

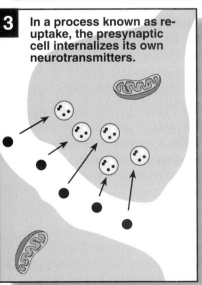

4 The postsynaptic cell reduces the number of available receptors for neurotransmitter binding.

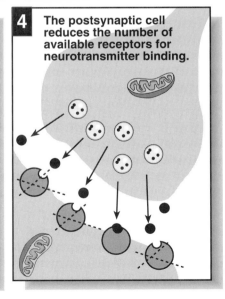

seaport. This hauling actually occurs in a process called reuptake, and is guided by molecules sensibly termed transporters. Once inside, the neurotransmitters are disassembled by still other molecules, proteins called monoamine oxidases. The net effect of all this activity is to reduce the number of norepinephrine molecules available for release.

These are just a few of the ways a presynaptic cell might change the effects of norepinephrine. But, just as commercial trade is a two-way shipping lane, the postsynaptic cell also has ways of altering the effects of neurotransmitter, effects that have the same depressive results as if norepinephrine was never shipped out. Here's how these work to create depressive illness.

You recall that the postsynaptic cells must have dock-like receptors on their surfaces in order for neurotransmitters to work. Another way to lower the effect of neurotransmitter is to alter the number of receptors capable of binding to norepinephrine on the postsynaptic cell. Using our analogy, let's say that the port of Zadar caught fire and most of its docks were destroyed. If a group of boats set sail from Ancona and reached the now burning city, they would not be able to dock. The ships might be forced to either dump their cargo into the ocean or turn around and go back to Ancona. In either event, communication between the two cities would be disrupted.

In some cases, it appears that something like a fire occurs on the surface of a postsynaptic cell in depressed individuals. Almost as if they had been burned, the number of receptors is reduced in certain postsynaptic tissues of the brain. If that's the case, then the norepinephrine molecules sent out by the presynaptic cell will have no place to go once they reach the other side. Communication is obviously disrupted, and the effects of the neurotransmitter negated as surely as if they had never been sent out.

That's not the whole story regarding receptor number, unfortunately. Some depressed individuals have *more* receptors on their postsynaptic cells than normal (even as they have less norepinephrine than is healthy). In certain patients, it has been discovered that postsynaptic cells do the opposite of setting their ports on fire. Rather, they add to their collection of docks if not enough norepinephrine is being received. In such patients, the analogy is that Zadar, sensing that there is not much shipping activity, starts to build docks both north and south of the city, trying to pick up any stray ship that might come along.

There would be no building activity if the amount of shipping activity were normal, of course. In a similar fashion, a postsynaptic cell struggles to keep normal activity in the presence of reduced neurotransmitter levels.

The receiving cell may actually increase the number of receptors on its surface, sensing that the normal amount of norepinephrine is not coming across the synapse.

As seen by these data, there are many complex processes involved even when one is attempting to explain the conceptually simple observation of lowered norepinephrine levels (see Figure 22 for a summary of these data). Unfortunately, things are about to get more complicated because norepinephrine is not the only neurotransmitter thought to be involved in human depression. I told you before that we were going to cover three brain systems before leaving our tour of the biology of depression. We have so far talked about two of them, the brain areas involving cortisol and the neural highways involving norepinephrine. The last neurotransmitter to discuss is now familiar to you – let's look at the role that serotonin plays in human depression.

The third brain system

As discussed in Chapter Six, many people believe that serotonin plays a role in human aggression. But this multi-talented neurotransmitter has many job descriptions in the human brain, and one of them is to play a role in human depression as well.

Most of what we know about serotonin and depression comes from a monumental historical accident, unusual in that it turned out to have a wonderfully happy ending. The accident involved a war and a drug company. As you know, Nazi Germany during World War II pock-marked much of Southern England with the V2 rocket. The first of these menacing machines were fueled by a combination of liquid oxygen and ethanol. As the war closed in on Germany, however, supply problems forced the rock-eteer-warriors to find another fuel. They found one, a chemical called hydrazine. When Germany was defeated, the allies discovered great stores of this nasty stuff (it is poisonous and caustic as well as explosive). Hydrazine soon became available at rock-bottom prices on the open market. Drug companies, ever on the alert for bargain organic compounds, bought large quantities of the fuel and started doing experiments.

One company, Hoffman–LaRoche, made a series of derivative chemicals from hydrazine that turned out to have powerful effects against, of all things, tuberculosis. And that's when the accident occurred. One of these chemicals, iproniazid, was found to have side-effects that made the patients taking the drug remarkably euphoric. Another chemical, isoniazid, was found to be a powerful inhibitor of monoamine oxidase (remember that neurotransmitter cannibalizer we just talked about?) and was found to have mood-altering effects. Purely by accident, the first antidepressants were discovered.

Only a few months later, a drug called imipramine was placed on the European markets. It was the first of a powerful class of antidepressant medications called tricyclics. The beginning of what some people term The Antidepressant Era had begun. The mood-altering – and in many cases life-saving – abilities of these medications have had marked effects on the way we view the brain, and a lasting impact on the field of psychiatry. (See Figure 23 for a description of the classes of antidepressant medications.)

More than forty years has passed since the first antidepressants were discovered, and there are now many such medications on the market today. These drugs have been of tremendous interest to researchers because of the powerful *biological* insights they bring to the nature of the disorder. And this is where the story of serotonin takes center stage. Many antidepressants work by changing the way certain neurons utilize serotonin, and that has led to what some researchers are calling the permissive hypothesis of depression. This hypothesis is described below.

Permission and the serotonergic pathway

You remember from our discussion of norepinephrine that many neurons linked together in a common pathway use a common neurotransmitter when they communicate. These highways are given names, such as "noradrenergic" for those neurons communicating with norepinephrine. However, there are also other highways in the brain that use different neurotransmitters to jockey information back and forth. One such highway utilizes serotonin and is called the serotonergic system.

If you were to stain the neurons involved in the serotonergic system, it would look like a spider web growing inside the brain. The neurons in these areas shoot out (the proper scientific term is "project") into an almost bewildering number of brain areas. Included in their purvey are some familiar regions, including the amygdala, hypothalamus and higher cortical areas responsible for some of our most sophisticated thinking. They even communicate with neurons in other neural pathways; for example, those involved in the noradrenergic system. This cross-talk of neural systems is vitally involved in the permissive hypothesis of depression, a fact we will return to in a moment.

Classes of antidepressants

Antidepressants can be categorized into three different classes.
What they are and how they work are summarized below.

A MEDICATION BY SERENDIPITY

Antidepressants were discovered by accident. Working with excess fuel chemicals left over from the Nazi-era V2 rocket program, Hoffman–LaRoche discovered the antidepressant properties of the drug iproniazid (originally investigated for its antituberculosis properties). A large number of antidepressants have since been discovered, classified as tricyclics, MAOIs and Second Generation Medications.

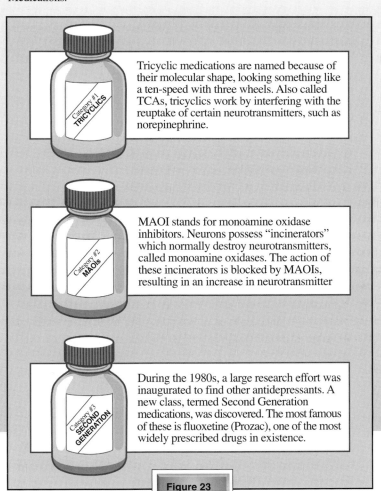

Tricyclic medications are named because of their molecular shape, looking something like a ten-speed with three wheels. Also called TCAs, tricyclics work by interfering with the reuptake of certain neurotransmitters, such as norepinephrine.

MAOI stands for monoamine oxidase inhibitors. Neurons possess "incinerators" which normally destroy neurotransmitters, called monoamine oxidases. The action of these incinerators is blocked by MAOIs, resulting in an increase in neurotransmitter

During the 1980s, a large research effort was inaugurated to find other antidepressants. A new class, termed Second Generation medications, was discovered. The most famous of these is fluoxetine (Prozac), one of the most widely prescribed drugs in existence.

Figure 23

How serotonin works with the permissive hypothesis

Knowledge of serotonin's involvement in human depression came primarily from studying the molecular mechanisms behind antidepressant medications. It was astonishing to many scientists that biochemicals that altered mood could actually save a person from suicide, and they were intensely curious to find out how these drugs worked.

Many antidepressants work on a very particular portion of the presynaptic/postsynaptic communication system between two neurons. In our previous discussions of norepinephrine, we described the cannibalizing properties of presynaptic neurons. A transporter molecule is fully capable of grabbing norepinephrine molecules in the synapse and stuffing them back into the presynaptic nerve cell of origin, in a process called reuptake. There is a similar reuptake process available in neurons that use serotonin. If serotonin is available in the synaptic space, a transporter may bind to it and bring it back into the presynaptic cell's cytosol. Many antidepressants focus on this reuptake process.

In many depressed individuals, it appears that the serotonin reuptake process gets too enthusiastic. The result is that too much neurotransmitter is taken out of the system, lowering the concentration of neurotransmitter in the synapse. As we have seen with norepinephrine, lowering such concentrations can prove disastrous for the person experiencing it, who then becomes depressed. In the case of serotonin, this is where the mechanisms of certain antidepressant medications take over. These medications have the ability to stop this over-enthusiastic recycling of serotonin. In fact, the newer drugs are often called SSRIs, which is short for Selective Serotonin Reuptake Inhibitors. The name of the class of drugs actually describes their biochemical method of action.

The lowering of serotonin levels is not done in a vacuum of course. As just mentioned, the serotonergic system is a neural highway with paths winding their ways throughout the brain, and touching many disparate regions. That spread brings them in close proximity to other neural pathways, including the noradrenergic system involving norepinephrine. This linkage caused a group of scientists in the United States and another group in England to form a hypothesis that joins the systems together. In their view, this over-enthusiastic serotonin recycling program causes a depletion of neurotransmitter throughout critical regions of the serotonergic system. This drop promotes, or, in their terms, "gives permission for" another system – the noradrenergic system – to drop levels of its workhorse neurotransmitter norepinephrine as well. In recognition of this association, they call their theory the permissive hypothesis of human depression.

So back to the haircloth

The permissive hypothesis allows for a number of scientific predictions to be tested under laboratory conditions. And while it makes for insightful explanations for a promising number of research efforts, the permissive model does not explain everything in human depression. For example, many depressed people have lowered numbers of serotonin-producing cells in their brains. Serotonin receptors (and there are at least thirteen different varieties!) have been shown to be greatly reduced in numbers in certain populations of depressed people. In fact, problems in receptor number may be the most important clue for explaining certain categories of depressive illness. A critical association has not been demonstrated between these systems and the CRF data we discussed several pages ago. This is the information linking cortisol and ACTH with perceptions of human stress. Indeed, there may not be just these three brain systems involved in depressive illness. (For example, we did not discuss one type of antidepressant, a class that works by changing the activity of monoamine oxidases, those enzymes involved in cannibalizing neurotransmitter spare parts.)

How do we resolve these questions? That's where our metaphor of the haircloth, the one we discussed at the very beginning of this Chapter, comes in handy. The poor citizens of the Second Cornice wore such haircloth garments. It was done as a form of punishment, for the very presence of the haircloth on a body could cause a lot of damage in medieval times. The haircloth rubbing into the person's skin might create an abrasion, even an open wound. Insects inhabiting the haircloth might either directly or indirectly infect the open wound. Without soap and water and antibiotics, those abrasions might turn into running sores and even life-threatening infections. It is not that the open wound and the insects are somehow related to each other. It is just that the presence of the haircloth in the thirteenth century gave both mechanisms an opportunity to cooperate in wounding a human being.

In thinking about the way these brain systems might work together, it is very clear that such a cooperative pathology may be in place in depression as well. Most researchers do not believe that there is a single cause of human depression. But several causes working together on groups of neurons might in fact give us symptoms deeply associated with depression.

Consider the case of Susan, the woman who sank into such an envious despair from not being able to have children. Perhaps her genetics predisposed her to certain types of depressive illness. The fault could be almost anything, from overeager transporters and monoamine oxidases, to not enough nascent neurotransmitters or receptors. Perhaps the way she was raised shaped her brain in such a fashion that she might be more susceptible to specific types of depression than the normal population. For whatever reason, it might be that she created a cognitive abrasion, just like a haircloth rubbing against skin might create a potential wound. That's not necessarily a bad thing; as long as she experienced no stressors capable of hurting her, she would not become depressed. But, like insects filling that haircloth, a powerful stressor might infest her life. That's exactly what happened to her, in the form of fertility issues. Her genetic "skin" might not have been able to stand the pressure of such cooperative insults. She might very well have experienced a cognitive "infection" and sunk into a major depression. Charles Nemeroff, a leading researcher in the field, has called such cooperation between genes and stressors the stress-diasthesis model of mood disorders. It presupposes an interaction between the environment (stress) and DNA (diasthesis).

This haircloth metaphor is not useful for all types of depressive illnesses, of course. Not everyone has predictive genetic family trees in the background, and neither were they all raised in abusive situations. Indeed, there are those depressive illnesses which occur out of the blue, seemingly without cause or prediction. There are also those dysthymic illnesses we discussed earlier, where a person seems to experience a low-level depression for most of their lives. Dysthymia occurs independently of a patient's background. No model presented here addresses bipolar disorder either, the disease characterized by tremendous alterations in mood and activity. These gaps in our understanding might at first seem to be a bit discouraging and, in terms of a complete answer, almost not worth writing about.

However, there is another way of looking at these models, a viewpoint which both excites and encourages our generation. Many people who have suffered depressive illness think that it is their "fault" that they go through such agonies. Loved ones witnessing family members going through one often think the same thing. The idea is that a personal failing somehow brought them into the condition, and if they could only pull themselves up by their bootstraps they would get better. This idea of a personal failing for a physical problem has a long tradition in the human family, stretching back to Dante's time (where infertility might be seen as God's curse, and depressive illness an "unclean" spirit) and beyond.

These days, the fact that a *chemical* might actually affect a *mood* in a particular direction has profound implications for us, very much as we discussed in Chapter Six. Biology follows the rules of chemistry and physics, and these rules must be taken into account when we discuss human behaviors and human emotions. We have learned this lesson for certain diseases. Nobody thinks of diabetes as a sinful, personal failing, for example. With certain kinds of diabetes, one simply takes insulin to ameliorate the symptoms. The same is true of depression, only here the "fault" is not in the pancreas, but in a specific region of the brain. Taking antidepressants to relieve the condition is not conceptually different to taking insulin. It is simply a scientific perspective on human disease, now applied to the brain. One of the reasons we are briefly reviewing the history of mind/brain debates, as well as attempting to define human consciousness, is to show just how revolutionary this scientific perspective is.

In the last part of this chapter, I would like to use such biological optimism to return to our discussion of the history of minds and brains, as well as introduce yet another ingredient in the definition of human consciousness. We shall end the chapter by re-examining the seemingly odd decision to talk about depression when exploring a cornice devoted to envy. As we discussed previously, we do it for a reason; the presence of subjective conscious feelings sometimes "obscures" our ability to understand emotions in a scientific fashion. Looking at the distance between envy and depression may be a very good example of this awkward tendency.

Back to minds and brains

The fact that a chemical might exist that could prevent someone from committing suicide is historically a novel idea, even if it seems familiar to us today. In the last chapter, we discussed the context from which the idea of an antidepressant medication might naturally flow: the notion that if mental illness came from dysfunctional chemistry, then the cure might come from chemically correcting the problem. In the last chapter, we talked about Kraepelin and Freud, both of whom attempted to organize observations about the mind and brain in their own particular ways. We especially emphasized the biological psychiatrists, whose views of the mind and brain might be summed up by an offhand comment by the brilliant French biologist Jacques Monod, "The cell is a machine. The animal is a machine. Man is a machine."

In 1949, Gilbert Ryle, the English philosopher from Oxford, coined a phrase that has actually stuck in the throats of many who talk about mind/brain issues today. He said that Descartes' dualism, and indeed any dogma embracing the vitalism of an immaterial mind controlling a material brain, was a "ghost in the machine." It has no place in science.

So is there anything *more* to be said about the mind/brain debate, especially when we think of envy and depression and the incredible magic of antidepressant medications? Some of the great present-day scientific minds seem to see the mind-as-machine as an open-and-shut case. The answer, however, is that there is still a great deal to say about the subject, and with some enthusiasm. As we will see, siding with the reductionists can actually lead to a very unreductionist mode of thinking. Just because we can now grow neurons in plastic dishes and clone genes for behaviors does not mean that we know everything. Even insights as good as Monod's cannot claim to see the entire picture, even if he is quick to pronounce a monolithic judgement over it. I lay the groundwork for this call-to-caution by relating a conversation I had with a student several years ago. Her name was Kim and she was talking about a lecture I had just given concerning a man named Donald O. Hebb. Dr Hebb was a famous research scientist who revolutionized our way of thinking about thinking.

"I am trying to understand your ghost in the machine ideas," Kim declared as she plopped down on the chair across from my desk. She had a clarinet case in her hand. As she sat down, she seemed to switch topics. "Didn't you say that Hebb believed in cell *assemblies*?" There was a tone of demand in her voice. I nodded my head yes.

Cell assemblies? The student was referring to one of Hebb's greatest contributions to understanding how the brain perceives the world. Hebb first proposed that any frequently repeated external stimulus would lead to the creation of a cooperative network of nerve cells. Hebb called these social groups cell assemblies. "You said it was a closed system, right?" I nodded again as she continued. "That means the nerves act like a unit, like a marching band. And that the unit exists because of the connections between the cells – those synapses – between the cell *assemblies*. Okay?"

I nodded a third time, but this time I interrupted her. "But there is a difference between a Hebbian synapse and a regular garden variety synapse," I said, putting on my best scholarly voice, "which is a question of sensitivities. If one neuron in an *assembly* fired, even if it was a weak firing, the rest in the group might fire as well. You see, they learned something together and acted as a group. A Hebbian synapse is a biased connection within the assembly." I then related something I had not said in the lecture, but that seemed to peak her interest. I told her that Hebb believed not just in biased synaptic connections joining single neurons, but in coordinated connections and responses from vast groups of cell assemblies. "It's something like your marching band idea. Entire sections talk and respond to each other, with the single purpose of conceptualizing something. Build the assemblies up together and you've got an explanation of thinking."

"So when I think of music I like – or my marching band music – I am just activating little cell *assemblies*? Is that all a mind is?" Now she didn't look puzzled or demanding. And I had to admit, I was taken aback by her question. But now she picked up her clarinet and looked to leave the room as hurriedly as she had come in. I said quickly, "Before you leave, you might be interested in a lesson from the old man's words." I grabbed a book off the shelf, and fumbled for a quote by Donald Hebb.

"The problem with understanding behavior is the problem of understanding the total activation of the nervous system, and vice versa." I quoted. She paused, "I don't understand that at all. But I'll be back." And with that comment, she left the room.

I did not know that this would be the first of a series of conversations with this student, one more of which I will relate next chapter. But her question was a good one, for she was wondering how individual component assemblies can bring about singular perceptions. And she was not the first to ask it. Indeed, the notions and hypotheses brought about by people like Donald Hebb and others folded themselves into a discipline many researchers call cognitive neuroscience. This new discipline brings together many ideas formerly considered to be unrelated, but which now seem to be wholly integratable. One collection of ideas is strictly cognitive, the mental processes of knowing or perceiving something. Another collection of ideas is strictly neuroscientific, attempting to understand those physical systems or subsystems that underlie cognitive processes. In many ways, the entire subject of this book is an exploration of the physical side of cognitive neuroscience, with a nod to the cognitive at the end of each chapter. To understand what all this has to do with mind/brain issues, let's briefly go through some characteristics of each of these collections of ideas. Then we will be in a better position to understand Hebb's contribution to the mind/brain arguments. Let's take the example of Susan, the woman who committed suicide, to illustrate the cognitive side of this discipline.

As you remember, Susan could become quite depressed whenever she thought about babies. But what did the concept of "baby" mean to Susan's brain? The ability to understand the encoding and processing of even this simple word is something for which cognitive scientists have a great deal of interest. Susan could consult at least three systems just thinking about the word: a visual one (what a baby looks like), an auditory one (what a baby sounds like), and a semantic one (what a baby *means* – which to her would include feelings of loss). These three groups of ideas, these underlying cognitive constructs behind the simple word "baby", sound complicated, and they really are. The great contribution of the cognitive side of neuroscience is that language is no longer seen as a single process. Rather, the representation of a word in the brain is now seen as involving many different codes, some of which might belong to the sight system, some to the auditory, and some to the semantic one. These might all be computed and integrated in different ways, all in an effort to make sense of four little letters. For poor Susan, the end results of these multiple ways of thinking always ended in tears.

The great contribution of Hebb and others was to put these completely cognitive constructs into the real world of the brain. We discussed one of his ideas, the notion that there are underlying networks of cell *assemblies*, those closed groups of neurons that my student was remembering. As the years have rolled by, it has become clear that Hebb was really onto something. Many of these cognitive structures actually have their own brain systems, closed-loop groups of neurons working together to create perceptions. This can be seen in dramatic fashion by looking at research into how laboratory animals see things, one component of the overall coding discussed in the previous paragraph. When one looks at individual cells inside monkey's brains, it is clear that the visual system becomes ordered into a series of "maps", each map responsible for a different aspect of an object being viewed. Monkeys contain at least thirty-four different maps; some are responsible for the color of the object being seen, some for the orientation, some for shading, some for liaison work, referencing the object to other similar looking objects seen previously, and so on. These represent a sophisticated form of cell assembly. Humans appear to have the same kinds of maps in their brains. There is even evidence that discrete cells in these maps receive their stimulation from specific neurons in the retina. This correlation allows the brain to maintain a correct spatial representation of the world outside the skull (the mapping is so specific that adjacent cells in the retina direct their input to adjacent cells in the brain – that way objects that are right next to each other in the visual world are right next to each in the neural world as well!).

As you can see, Hebb had a great deal to contribute to the arguments of minds and brains. He gave us a framework whereby we might pin our cognitive ideas onto honest-to-God synapses, and, in so doing, gave us one of the great strides of twentieth century neurobiology. Kraepelin would have been proud, for Hebb seems to have placed the mind/brain arguments under the insightful glare of the scientist's microscope. And as we shall see in the next chapter – perhaps to Kraepelin's chagrin – the end result is that the argument jumped right off the slide.

Consciousness

It is clear from the previous paragraph that cognitive neuroscience is the reductionist's answer to getting us to think about mind functions in serious scientific terms. However, we are not just talking in general terms about the presence of cellular assemblies in this chapter, we are discussing specific human behaviors with real life consequences. How can phenomena such as envy and depression help us better understand the inner workings of the nerves beneath our skull? How does jealousy inform us about cell assemblies? Or human emotion?

As you recall, one of the ways to address this question is by exploring the model set forth by Joseph LeDoux, namely that subjective emotions arise when we become consciously aware that an emotional subsystem in the brain has been tickled. We have spent most of this chapter talking about several specific emotional subsystems involved in depression (now *there's* a group of cellular assemblies!), and we have been listing ingredients that make up the notion of consciousness all book long. We have discovered that consciousness incorporates functions such as memory, including long-term storage devices, short-term buffers and the neural go-between, the presence of working memory. We have found that there are areas in the brain responsible for our conscious awareness. We have also uncovered something perhaps surprising, namely that the creation of consciousness has neural connections with areas that are not within our awareness, but that are nonetheless important for maintaining consciousness.

I wish to add yet another ingredient to our recipe for consciousness, something that our discussions of envy, depression and Hebb can help clarify. Specifically I want to say that there are probably different *types* of consciousness, and people can experience varieties of them for different reasons. This idea of types is the next ingredient in our discussion. What do I mean by types? Many researchers believe that there are differences between conscious emotional feelings and conscious thoughts, even though both are involved in the idea of "awareness". Moreover, *emotions* like depression and *thoughts* like jealousy have to be understood in terms of these differences. That may sound a bit confusing, although Hebb would recognize this form of separation in specific closed-group terms, so let me explain.

At first glance, notions like conscious emotional feelings and conscious thoughts seem quite similar. Both involve many of the same ingredients in the consciousness recipe: working memory, short-term buffers, long-term storage devices, even subsymbolic systems that tend to work just below our level of awareness. The difference between them concerns some of those subsymbolic processes. Many scientists think that conscious emotions and conscious thoughts are generated by different systems. Not only do they involve different systems, there are different numbers of reactive processes between the two. It appears that emotional systems are much more complex, involving many more collections of brain systems than mere thoughts.

Let's use our example with Susan again to illustrate this point. Hanging around in Susan's brain is the objective perception that, while other couples can bear children, Susan cannot. There's nothing complicated about that fact (if a thought can ever be said to be "uncomplicated"); it is simply a realization in her brain. But also clanging around in Susan's brain is the emotional reaction to the fact that she cannot bear children whereas other couples can. Susan's emotional reaction is cataclysmic and complex; when she is in the midst of her recoil, many powerful subsystems in her brain become activated, as if the neurons had just received crippling separate shocks. As a result of some deregulation of those complex reactions, Susan then slipped into a depression that actually killed her.

The complexity of these reactions is underscored when we consider the fact that most researchers believe the following about depression: it is the dysfunctioning of a normal series of deep-seated responses to threats. It is the dysfunctioning of a *survival* system. Even though Susan's life ended in suicide, a great deal of her brain energy was marshaled as a survival response. And that's the whole point. Emotions, unlike thoughts, create a whirlwind of activity, and, unless a thought triggers a reaction, there is a great deal of difference between the two types of awareness. This idea, that there are different categories of consciousness, is the next ingredient we must add to our overall definition of the phenomenon.

So what can we conclude?

The idea that there are different kinds of consciousness helps us on the road to understanding how emotions and thoughts integrate themselves in the great mind/brain debates of the twentieth century. Perhaps a future Hebb will give us the exact cellular assemblies devoted to objective realization, even as others are working out the neurons involved in the complex emotional responses. But the presence of these separable systems also underscores a gap in our knowledge, a rift we actually discussed in the first part of the chapter. We started by saying that an association between envy and depression might seem a bit far-fetched. We also said that the fact of its oddness illustrates an important fact of emotions. I would like to discuss this important fact by way of summary before we leave this chapter.

We have seen that there are at least three emotion systems involved in depression. One concerns hormones such as cortisol and ACTH. The other two involve neurotransmitters, norepinephrine and serotonin, coursing through discrete neural pathways which may themselves be linked. Omitted from the discussion was the idea that emotion systems are devoted exclusively to invidious feelings. I left them out for a straightforward reason mentioned at the start of this chapter. No one has ever discovered even one emotion system devoted exclusively to envy. Indeed, professional researchers who have studied envy (perhaps seeking to find such systems and molecules) end up with genes not for jealousy, but for aggression, fear, sexuality, and depression. In other words, they find no such representation of the emotion of jealousy in the brain *per se*. There are only powerful underlying associations. This gap between the human category and the neurological reality is the phenomenon I wish to emphasize as we close this chapter.

The feelings that we sense our emotions to be sometimes distract us in our attempts to study them scientifically. I take it as a fundamental principle that the brain is primarily a survival organ. Brains must respond to certain parts of our environments in order for the human to compete successfully, and when the response is deregulated (as in depression), it is important to understand exactly what is going wrong. If we experience the brain's surface responses emotionally and give that surface reaction a category (like jealousy), that's okay – as long as we remember that we are describing *only* a surface notion. But if we don't *feel* the brain's responses, don't experience a subjective sense of the reaction, that is still okay. It is not the category of the feelings, or even the presence of the feelings, that is the essence of survival. Instead, the reaction of our brains to our hostile outer environments as evolution shaped us is where natural selection takes place, and where neurological reality lies. The feelings may simply be a by-product. The selective forces may not exert their effects at the surface of our awareness, or indeed look anything like the surface feeling being experienced. I am not at all surprised that certain types of depression, for example, turned out simply to be a crippled threat response.

This separation, much like understanding the different subsystems separating conscious emotional reactions from conscious thoughts, is not a trivial notion. The process that detects and interprets the environmental stimulus points us to the fundamental system, and is the reason why the brain is still hanging around after all these millions of years. That's true whether we have felt jealousy for millions of years or have just decided to create the classification in the last few thousand.

That's what the link – or lack thereof – between invidiousness and depression may tell us. You don't take anti-jealousy pills if you are crippled with envy. Such medications just don't exist. If you want to get rid of some of the negative aspects of jealousy, you may have to take an antidepressant. Knowing this idea – that generated feelings can sometimes get in the way of understanding real live brain systems – is not a bad way to look at human envy. Indeed, it's not a bad way to look at most of the behaviors mentioned in this book.

CHAPTER EIGHT

Pride

"... those that crawled along that painful track
Were more or less distorted, each one bent
According to the burden on his back

Yet even the most patient, wracked and sore,
Seemed to be groaning: 'I can bear no more!'"

-Canto X, The Purgatorio

"If God put accountants in charge of creating torture devices, they might have come up with something like the First Cornice punishments," the professor said dryly. By now we understood that our mentor liked to begin his lectures with some attention-getting statement. In this case he was talking about inmates interred in the level of Pride, the First Cornice, where proud people suffered their just deserts.

He continued, "When Dante and Virgil entered this level, they first beheld three beautiful bas-reliefs, hewn in marble and placed in an inner cliff face. They also noticed the inmates – who could help it? – shouting in despair, crawling round and round the cornice. Each shade carried crushing slabs of rock, not heavy enough to kill them, but heavy enough to elicit their agonies and cries."

The professor explained that each soul carried an amount of rock in direct proportion to the amount of pride they exhibited while alive. "That's what I mean when I say accountants could have designed this," he continued, "everyone was getting their punishments in a measured response. This punishment was easily the harshest the sojourners encountered in Purgatory, probably reflecting prevailing medieval attitudes about the serious nature of the great sin of pride."

The professor said that Dante and his friends heard descriptions of several historical incidences, each serving as an example of the problems associated with the folly of human pride. One of these incidences came from what was for Dante recent Italian history, the military defeat of the Aldobradesco family by yet another army from Siena. There were also examples of pride taken from Greek mythology, specifically discussing the gifted seamstress Arachne.

"Who did what?" the red-haired boy interrupted, raising his hand as he spoke, as if he were the only one in the room.

"She attempted to best the weaving prowess of the goddess Athena," the professor explained, patient with the interruption. "But that was not the best story. The final one involved Tomyria, the ancient queen of the Scythians, who successfully fought a Persian emperor but became too haughty in her victory. The point was straightforward. It was to remind the medieval reader that God was not tolerant of rivals."

The poets were not allowed to leave this level until they encountered the Angel of Humility, as if to emphasize the point. After receiving a blessing (and, for Dante, even a touch), the poets scrambled up a narrow path and were gone.

The subject of this chapter

The Biblical sin of pride was no small issue in the thirteenth and fourteenth centuries. One can observe the emphasis by seeing how harshly Dante imagines their punishment. Pride, as you might suspect, is also going to be the subject of this chapter – or at least, part of the subject of this chapter. Once again I have to admit to a certain amount of frustration concerning the science of the subject. How does one discuss the *biology* of pride? Like avarice and envy, no one has ever isolated a gene responsible for human egotism. No one has ever identified a region of the brain that is exclusively devoted to boastful feelings. We are forced to deal with something that appears to be a part of human nature yet may possess no biological correlates worth discussing.

It turns out that our task may be a little less formidable than that in previous chapters. In Dante's world, pride was considered, like other sins, to be a fault of the human personality. And while we have addressed emotional systems that govern specific characteristics of our personalities, we have never tackled the subject of the "self" from a genetic point of view. It's not that we left the subject alone, of course. Self-awareness is what consciousness is all about, which is a great component in any definition of self. But we've discovered that consciousness is such a complex notion that we do well if we can simply mention a few necessary ingredients. And we've completely ignored any genes that might be involved in the process.

If a study of egotism begins with a study of self-awareness, how do we isolate the genes for self? The data will lie, if they exist at all, somewhere in the brain. Somewhat reminiscent of Descartes, the answer to the question, "Who am I *really*?" comes down to, "Your identity is whatever the brain thinks you are." At its most fundamental level, then, identity and consciousness come down to the presence or absence of thought. And that gives us the clue to the subject of this chapter.

Neurobiologists are learning a lot about the neural basis of thought. Even if we can't discover a gene for conceit, we might be able to explore the biology of thinking, and then make a few guesses about how self is formed, perhaps even how ego is created. Since we discovered that a large part of awareness involves recalling and responding to various kinds of memories, we can begin to explore the genetic correlates of storing information, especially as it relates to our own self-perception. How then do neurons help us in this process? Indeed, how do neurons learn anything?

The goal of this chapter is to explore at a genetic level the processes just mentioned, memory and learning. We will start with some general comments about the formation of identity in a human being, talking about an outdated debate that used to be called nature versus nurture. Then we will describe a carousel of some of the most interesting genes ever discovered. These sequences are directly involved in the process of wiring up the brain while the baby is in the womb, and then teaching the newly minted brain certain things once birth has occurred. We will end our discussion in the usual manner, by revisiting our mind/brain discussion, then discussing a final aspect of human consciousness.

In many ways, this subject serves as a summary of everything we have mentioned in the previous chapters. Lust, gluttony, wrath, greed, fear, and envy all depend upon the ability both to create and remember specific thoughts. The sum total of these emotion systems forms major parts of who we are. Understanding how memories are created, and indeed how anything is learned, is the final, most intimate subject of the Genetic Inferno. It is our last task to address the genes that make it work.

An argument out of fashion

I just mentioned that I thought nature versus nurture debates, the questions surrounding whether humans are products of their birth or their environments, are outdated. This rather pejorative judgement deserves some explanation. For the past several decades, it has been fashionable for many researchers to think of the human brain as a "tabula rasa", a blank slate upon which the environment can write almost anything. This notion emphasizes the nurture side of the equation, the belief that human beings are a strict product of their environments, with genetics playing a secondary role (if indeed any role at all). Some of this emphasis is understandable. At the beginning of the twentieth century, the so-called eugenics movement was in vogue, which gave great weight to inherited characteristics, the nature side of the argument. Extensions of this notion included the idea of "proper breeding", which in turn gave rise to some of the most horrible crimes ever committed on the face of the Earth.

Like a teenager who has suddenly grown up, we are learning that the extremes, while valid in exploration, do not reflect the final reality. People with their confounding personalities are products of both nature and nurture. This balance starts very early. From the moment the first cells are dividing, a human embryo is learning how to distinguish self from nonself. Neural connections help it to discover where its arms are, where its nose is, that its head is separate from its feet, and so on. Cells sensitive to touch become hardwired into an area of the brain known as the somatosensory cortex. The embryo actually forms a tiny image of itself, which some researchers call a homunculus (literally "tiny man"). You can find in many textbooks a picture of this tiny man, usually drawn as a distorted cartoon image of a human. Those areas most exquisitely sensitive to touch, hands, genitals, specific areas of the face, are represented in the drawing as much larger features than those areas that are less sensitive.

The ability to create the homunculus is hardwired into human DNA. By that I mean there are genetic processes that govern it, a few of which we will explore in this chapter. But while this neural circuitry is being soldered into place, another equally extraordinary process is occurring. Genetic processes are molding specific neurons to become sensitive to environmental inputs. That is, they are hardwiring neurons to *not* be hardwired, meaning they have the capacity to learn from the outer environment. One of the most dramatic examples of this lies in the creation of human vision. If a child is born with cataracts – inhibiting light from coming into the eyes – and the cataracts are not removed within six weeks after birth, the child may be permanently blinded. What's so amazing is that the rest of the visual apparatus may be perfectly intact. The neurons for vision, and all the accessory organs may be correctly formed, ready to see objects, ready to communicate vision. But if the cataract prevents light from entering the eyes to stimulate neurons throughout the visual circuitry, certain connections will not fine tune themselves. Without the external stimulus, they may not even form. This event is so critical to development that it is even timed; if light does not come within several weeks, the baby's neurons will lose the ability to *ever* connect themselves. The poor child may become blind as a result, even if the cataract condition is corrected later in life. In the parlance of neurologists, the ability to see is "experience dependent".

This is a powerful analogy for showing how both nature and nurture work together to create a functioning human brain. Without the genetic hardwiring, the baby wouldn't even create an eye. But without the environment, the baby may never acquire vision. Most researchers believe that human personality is no different than vision in this regard, though the cues are much more subtle. Human personality, including the self-confidence so punishable in the First Cornice, has both genetic and environmental roots. As more and more relevant genes are isolated, this more synthetic view gains increased support. That's why I can use words like "outdated" to suggest that only one side of the nature versus nurture debate is the true explanation of human personality. The real research trick is to find out the relative contributions of each.

A road map for the chapter

In an attempt to understand how brains develop and learn things chapter describes a wide variety of gene systems. Here's a map may help to keep all the sequences straight.

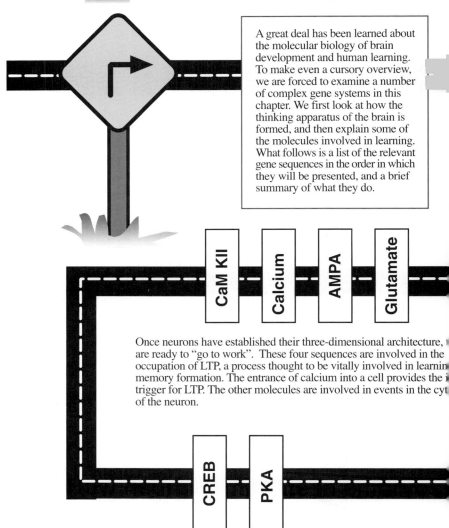

A great deal has been learned about the molecular biology of brain development and human learning. To make even a cursory overview, we are forced to examine a number of complex gene systems in this chapter. We first look at how the thinking apparatus of the brain is formed, and then explain some of the molecules involved in learning. What follows is a list of the relevant gene sequences in the order in which they will be presented, and a brief summary of what they do.

CaM KII — Calcium — AMPA — Glutamate

Once neurons have established their three-dimensional architecture, are ready to "go to work". These four sequences are involved in the occupation of LTP, a process thought to be vitally involved in learnin memory formation. The entrance of calcium into a cell provides the trigger for LTP. The other molecules are involved in events in the cyt of the neuron.

CREB — PKA

As the neuron "learns" something, events occur in the nucleus that are involved in gene activation. One of the most important gene products in m formation is CREB, working in association with a protein called PKA.

Figure 24

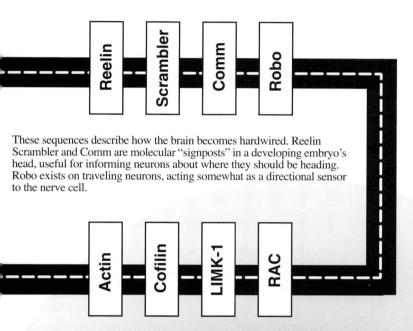

These sequences describe how the brain becomes hardwired. Reelin Scrambler and Comm are molecular "signposts" in a developing embryo's head, useful for informing neurons about where they should be heading. Robo exists on traveling neurons, acting somewhat as a directional sensor to the nerve cell.

These sequences are also involved in the molecular and cellular construction of the brain. A mutation in one of these genes is responsible for a certain kind of cognitive defect in humans. These sequences thus provide linkage between molecular development in embryos and adult thinking processes.

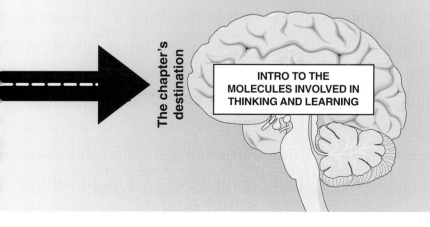

The chapter's destination

INTRO TO THE MOLECULES INVOLVED IN THINKING AND LEARNING

Given that this is so, and that both memory and learning make up the bulk of who we are, how does the brain go about wiring itself together in such a fashion that the concept of "self" can be established? And once wired, what are the genes that allow this brain to be so plastic that it can actually respond to environmental cues? Obviously, these questions are really asking for the genetic contents of both nature and nurture. We are going to explore a few genes from each category, looking at the sequences involved in wiring the brain together and at those involved in helping us learn things once the wiring has taken place. This is enormously complicated stuff, and we are going to talk about no less than twelve different gene sequences. Feel free to use Figure 24 as a roadmap if the jargon gets confusing. Even with this complexity, the journey will ultimately be unsatisfying; as you will see, we have a long way before we understand how brains make behaviors, a frustration we have visited in every chapter so far.

I would like to begin as we usually do when something seems a bit convoluted, with an analogy taken from some activity in the thirteenth century. In this case, the activity comes from a particular historical figure, a man Dante actually meets midway through his journey through the First Cornice.

A battle of pride

The man's name was a tongue twister, Omberto Aldobrandesco, Count of Santafiora. Dante found him in a very repentant mood, sorrowful for the pride he experienced in life. According to Dante, the district over which the Count resided was a lawless land, with Omberto's famous father, Guglielmo, its chief robber-baron. When his father died, Omberto took up his father's evil ways, becoming famous mostly for being proud of his family's checkered ancestry. This display of great pride, especially in the context of great evil, is why Omberto ended up an inmate in Purgatory's First Cornice.

The Count administered his corruption from a castle located in a town called Campagnatico. As you might expect from such notoriety in reputation, Campagnatico was at odds with just about every one of its neighbors, including much of the leadership of the nearby town of Siena. Sick of it all, the Sienese decided to march out to Campagnatico in 1259, lay siege to the castle and then lay claim to the lands of the Count. The Sienese did indeed march out to the castle with a large army, much greater in numbers than Omberto could muster. That meant absolutely nothing to the proud Count, who refused to surrender, instead trundling out his meager force to meet the enemy. It was a blood bath, though surprisingly for both sides. The Sienese army closed around the Count's forces, who immediately charged into the center of the opposing army, killing many of the invaders. But his efforts, though valiant, were ultimately quite futile. Many men from Siena died that day, but so did soldiers of the Count, including Omberto himself. With the proud head destroyed, Siena conquered Campagnatico, took over the district and broke the back of the once powerful Aldobrandesco family.

The relevance to brain development

The ugly fact of thirteenth century Tuscany was that might often meant right. Even though their large army suffered severe losses, Siena won the day because of overwhelming force. But why talk about such history in a chapter on brain development? This migration to a place where severe losses still end in a claim on territory, such as happened to the Sienese army, is remarkably similar to migrating neurons and brain development. One of the most interesting discoveries in developmental neurobiology is the amount of neural warfare that precedes the construction of individual brain regions. It might surprise you to know that many neurons die before the proper connections are established.

Here's how it works. When it is time for neural structures to begin appearing in a developing embryo, one immediately notices an abundant supply – perhaps *over*abundant supply – of neurons throughout the tiny body. These plentiful nerve cells migrate to various areas of the developing embryo as if marching off to war. What forces guide these neurons to migrate to specific places remain a mystery, though we have recently begun to get a handle on some of the genes involved in the process (we'll discuss one of them in a moment).

The upshot is that numbers of neurons begin marching like an army across the cellular real estate of the embryo to various locations. Some locations appear to have too many invaders for the developing baby's satisfaction. In response to this overcrowding – here's what is so surprising – some of the neurons in a given location begin to die off, almost as if they were being slaughtered in a battle. The carnage can be massive. In some places the kill-rate is as high as 80%. Some places are a little less violent, suffering only 20% casualties (how many actually die depends upon the location one is observing in the embryo). This warfare is not a trivial event, nor is it accidental, aimless or random. Data show that this death must occur in order for brain territory to be "won", that is, to be organized into a functional configuration useful to the adult. This is much like Siena's army migrating to a specific area and then, after significant losses, annexing a territory.

But that's not the end of the surprises

From a research point of view, the obvious key to understanding brain development lies partly in figuring out why these neural cells die. In recent years, scientists have uncovered the mechanism of cell death, and in so doing uncovered an equal surprise. It turns out that many developing neurons die not because they are murdered, *but because they commit suicide.* You did not read that wrong. A certain percentage of nerve cells in the developing brain migrate to a given area, pull out their molecular swords and fall on them. It is a deliberate death, a form of self-immolation scientists have come to call apoptosis. Under the microscope, you can actually watch a neural cell shrivel to nothing in three discrete steps.

1. The chromosomal complement of a cell (called the chromatin) condenses inside the nucleus.
2. The outer membrane of the cell loses structural integrity. One observes protuberances, or "blebs", extruding from the surface of the cell.
3. Like a garment at the wrong water temperature, the cell begins to shrink. In the nucleus, the chromosomes start to fragment, as if someone has taken scissors and cut the DNA into small pieces. The cell dies (see Figure 25).

This cellular suicide is so important to the developing embryo that if you prevent it from occurring, you also prevent the brain from forming correctly. But why does a neuron decide to die at a particular time and in a particular space? Why do groups of neurons migrate to a specific area at all, if only to commit suicide once they arrive?

The answers, unfortunately, are as mysterious as the questions are obvious. However, we are beginning to get a rudimentary handle on some of the migration mechanisms, and the view hints at a powerful idea: specific kinds of neurons are *attracted* to specific areas within the developing brain.

How cells commit suicide

Creating the neural architecture of the human brain involves much deliberate cell death. The process is termed apoptosis.

NERVE CELL DEATH

The process of apoptosis occurs in a specific series of steps, as shown below. Changes are first seen in the chromatin (a term meaning both chromosomes and associated molecules). Later, alterations in the outer physical appearance of the cell are observed. Eventually, the cell dies.

A The cell's chromatin begins to condense. Individual strands of DNA (chromosomal material) are observed in the nucleus.

B The outside membrane of the cell loses its structural integrity. Protuberances called "blebs" are observed on the cell surface.

C The physical size of the cell begins to shrink. The condensed DNA becomes fragmented, destroying gene function. The cell eventually dies.

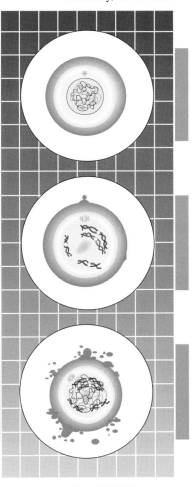

Figure 25

A great deal of evidence suggests that neurons have the ability to sense (1) where they are going (at least they know when to keep migrating past certain regions in the cellular landscape), and (2) when they should stop moving, knowing that they have finally arrived at their final destination. This implies the existence of tiny signposts in the brain, directing traffic, creating routes the neurons may recognize as they seek to discover their resting places. I say routes because neurons stream in from all areas of the embryo in an attempt to wire themselves correctly to the developing head. This idea is a bit complicated, so let us return to our analogy of the Sienese army marching out to Campagnatico, and use a thought experiment to clarify a few points.

Suppose some dark angel of Dante's imagination decided to play a nasty joke on the Sienese army as they set out to battle against Omberto. Suppose the angel magically whisked twenty companies of men away from their Northern Italy homelands and transplanted them to the beaches all around the Southern coast of Italy's boot. If Siena still wanted the Count's lands, these companies would be forced to march northward and attempt to find the battlefield. In thirteenth century Italy, without the benefit of satellites or even good maps, this would be no easy task. The various companies would be forced to seek signposts, use available roads, ask questions of people in the towns they encountered, even look to the traveling experiences of individual members, all in an attempt to find Campagnatico. From a bird's eye view, you would see vast companies of soldiers streaming from the beaches of Southern Italy into the interior, marching deliberately from one place to the next, moving in fits and starts, seeking directions, moving ever northward. Given enough time, they might all converge on Campagnatico and fight their war.

That vision of individual members streaming to a particular location is what many scientists think actually occurs during human brain development. Neurons from both local and distant locations start migrating to specific areas, seeking signals, using cellular "highways" complete with molecular signposts, moving equally in fits and starts. Given enough time, they converge at their destinations, and the neural slaughter commences.

Recently, a number of the neural "highways" and genetic "signposts" that young neurons use to find their destinations have been discovered. When one gene, called *reelin*, is inactivated in mice the resulting animals have parts of their brains scrambled in critical areas. This scrambling results in the mouse walking with a kind of staggered, reeling gait, hence the name. When the researchers looked at the scrambled regions of the brain affected by this lack of *reelin*, they found that many migrating neurons had not reached their proper destinations. Instead of finding intricately wired connections, the scientists observed massive pile-ups of neurons that appeared not to know where to go. They also discovered that the product of the *reelin* gene, the Reelin protein, is actually expressed in areas of the brain to which these scrambled neurons are supposed to hook-up. In other words, the Reelin protein acts like a signpost. This signpost advertises to specific neurons that the cells are supposed to migrate to a specific area. Without the signpost, the neurons never arrive, piling up in specific, unorganized clumps (or if they do arrive, they keep moving, never settling in a proper arrangement). Without the proper arrangement, the brains of these mice become scrambled and the gait, amongst other neurological deficits, is created.

The second gene comes from a mutation called *scrambler*. The lack of this gene also results in neurological deficits that can be observed in the way mice walk. One can also observe the same massive pile-up of neurons that seems unable to connect to anything. Most confusing (and exciting) to researchers is the fact that this gene is located on a different chromosome than the *reelin* gene. This *scrambler* sequence now goes by the confusing name *mDab1*, which is short for mouse disabled homologue 1, to distinguish it from *reelin*. And what does *mDab1* do? Whereas the Reelin protein is made in localized regions of the brain, acting like an informative signpost, the *mDab1* gene is made in those piled-up neurons (the neurons that are supposed to hook-up to the areas designated by Reelin). mDab1 protein may be acting like the eyes of those soldiers scattered around the Italian countryside, continually scanning the roads for signposts back to Campagnatico. In other words, mDab1 may be acting as a specific sensor for the Reelin protein. And that makes sense. Without mDab1, one sees a familiar lack of neural organization within the brain, correlated with defective motor skills. Scientists believe that they have actually found a mechanism that may shed a great deal of light on neural migration.

What this has to do with our discussion

The story of *reelin* and *scrambler* is an exciting one, and for a couple of reasons. The first has to do with what these mutations *don't* say: even though they are important for brain development, the absence of these genes doesn't fully cripple the development of the brain. Many connections are still forged correctly. Indeed, it's not as if these mutations keep the mouse from walking, a very complicated thing to do – the mice just can't walk well. Why is that exciting? It means there are many different functions that *reeler* and *scrambler* don't mediate. There must be other signposts marking specific brain areas, and probably many different sensors scattered throughout neurons in the developing brain interacting with them. The analogy is the dark angel taking the armies from perhaps thousands of different conflicts in thirteenth century Europe (and there *were* thousands) and flinging the soldiers to beaches across the continent; they would be forced to stream in from all directions, migrating across huge distances in many different countries, in order to arrive at their designated battlefields.

There is another reason why these results are so exciting. The interactions of these proteins gives us one of our best views of the molecular hardwiring of the brain. The *hardwiring* of the brain, the genetic basis for brain function, the permanent circuit board of thinking. Since we are defining our sense of self as an interaction between unchanging genes and constantly changing environments, *reeler* and *scrambler* provide us with insights into the sorry-you're-born-with-it side of the phenomenon. If the genes are functioning normally, the creature has a chance of interacting with its environment in a successful way. But if these genes are mutated, the creature is rendered dysfunctional, regardless of what environment surrounds it.

What about other molecules?

Reeler and Scrambler give us great insights into the migration and settling of neurons as an embryo grows. But this idea that some molecules serve as an attractive "stop here" signal for certain neurons may not be the whole story. There could also be signals, for example, that say in essence "wrong way, stupid" to neurons that are trying to settle down in the wrong place. Such signals might result in the death of the neuron itself, or might simply force the slithering neuron to reroute itself. Thus, there could be several kinds of information that a neuron responds to as it seeks its final architectural configuration.

There is strong experimental evidence to support this notion of a single neuron responding to multiple signals. A number of genes have been discovered in fruit flies that give neurons a choice of guidance options. These genes have also been found in humans, but since their function has been better worked out in the insects, we will talk about these first. One must be careful, of course, and the usual caveats about nonhuman to human leaps in logic apply here as they have in other chapters. To understand a little about how these genes work, we need to talk about groups of molecules we have not yet discussed, and then move to the genes inside specific neurons.

The forest in the brain

It is tempting to think of the neural architecture of the brain as a vast forested valley. Indeed, one readily sees neurons as large tree-like structures under the microscope, each trunk equipped with intricate twig-like structures, branching into increasingly smaller and smaller whisps of tissue. The neurological canopy created by this forest appears both breathtaking to behold and so indescribably complex that designing experiments to understand it seems like an impossibility. In many ways, it truly is.

Even this forest analogy isn't as complex as the real-world situation of the human brain. At the cellular level, the brain actually looks like someone took a giant airplane and dropped tons of dirt on the forest, filling it to the top of the highest tree. This "dirt" is called the extracellular matrix, and all neural cells are embedded within its semi-solid framework. Understanding how neurons work means gaining an appreciation of the matrix in which they are immersed.

This interactive matrix is not made of random land-fill. Rather it contains complex molecules, many deliberately placed in regions of the brain to affect certain functions. The dirt piles not only play a role in normal adult cell functioning, they also have a great deal to do with developing embryos, including the neurons that create the brain and hook-up the organ to various neurons in the body. We now look at how the "dirt pile" works to guide the creation of our nervous system.

A worm with the head of a computer

Developing neurons are interesting cell types to behold. They look less like trees and more like nematodes inhabiting the tree's soil. They move like worms too, slithering along the insides of the embryo's body. When one examines the growing tip of these developing neurons, one finds a fascinating structure called the axonal growth cone. I use the word fascinating because of the growth cone's function; this tip has a powerful "intelligence", composed of so many different sensors that researchers think of it these days like a tiny computer. There's a mixed metaphor for you, a worm with the head of a computer. As the embryo develops, more and more neurons send out their intelligent axonal growth cones, guiding neurons to specific places, eventually forming a brain and nervous system.

So, just what does this computer sense? The growth cone processes many different kinds of signals, including those involved in directional migration. Directional cues are really quite simple: the worms are either attracted *to* a certain place (like the Reeler phenomenon we just discussed) or are repulsed *away* from a certain place (which we haven't talked about yet). This series of stop/go cues helps inform the developing neuron as to where it should eventually settle as it migrates inside an embryo.

At the molecular level, the reasons for such location discrimination are beginning to be understood. It is a complex interaction between molecules embedded in the developing embryo and receptors that are tethered to the surface of the growth cone. Many of these embedded molecules exist in the extracellular matrix, which is why understanding the composition of this "dirt" is as important as knowing how the "worms" crawl through it. It turns out that developing embryos sprinkle different areas of their bodies with different components in their matrices. And different neurons display different sensor molecules in their growth cones, some capable of sensing attractions/repulsions in one kind of matrix but not another.

The components of the computer

Most of these sensors are receptors, which, if occupied, transduce signals down to a neural nucleus and relay to it specific kinds of information. Creating a nervous system means having to picture millions of neural cells slithering around a developing organism, assaying the external environment, and deciding where to go based on what they encounter. This two-dimensional decision-making protocol – neural tip interacting with extracellular environment – is a hallmark of neural development.

I would be painting too simple a picture, however, if I said that when a receptor is occupied, the neuron immediately starts responding in a specific, monolithic kind of way. Biology is always messy, and even this interaction is not simple. For example, cells will often respond differently to the number of receptors being occupied at any one given time. This is convenient because the external growth factors are often encountered in a gradient fashion, with concentration providing the directional cues a given cell must sense (a cell might be attracted initially to a few molecules binding to a few receptors, for example, with the attraction increasing as more receptors are occupied). That means that, over time, a different number of external molecules will be encountered depending upon the location of the neuron. In the end, the positional information the neuron receives may be concentration dependent, and different neurons will stop/go at different times depending on the concentrations they sense from many molecules. How messy indeed!

What our studies of fruit flies have told us about humans

Such messiness requires most of us in our laboratories to look first at simple creatures before moving on to more complex ones. Indeed, researchers didn't start out cloning human genes in an attempt to understand how neurons become wired together in a bilateral fashion. Old standby workhorses, like fruit flies, were characterized first, and remain the developmental organism of choice to this day.

It may seem odd to turn to an insect and expect to find clues as to human development; indeed, if no human relevance had been discovered, I doubt the work on insect biology would continue to be funded as heavily as it is today. What has astonished a lot of us is the amazing conservation of structure and function displayed by neural cells. Even though there is a great deal of evolutionary distance between arthropods and hominids, you won't find the gap nearly as daunting when you look at the DNA. Many genes that work in the nerves of fruit flies have human neural equivalents, and are similar not just in structure, but also in function.

One of the structures common to both fruit flies and ourselves is an embryonic edifice known as the midline. This is a combination of structures – groups of cells, extracellular matrix, embedded molecules – that literally divides the embryo into two halves. This is the molecular derivation of bilateral symmetry, and has much to do with how the embryo gets its two sides, and even how the embryo gets its head in the right place. You can actually see the midline in most developing embryos.

Both human and insect embryos pay a great deal of attention to this midline. The reason is that it is chock full of molecular signals, sending out an incredible number of growth cues to its tissues. Neural cells are some of the tissues deeply affected by molecules in the midline. You can actually watch the axon growth cones of local neural cells send out extensions to the midline, cross it, immediately turn left, and never cross the midline again. Alternatively, they turn right, and show the same permanent no-turning back, or they do not cross at all, but instead run away like

How neurons migrate in embryos

To create neural structures in embryos, neurons migrate throughout the developing body. Where they go and how they connect involve processes of attraction and repulsion. Here are some of the signals they use to make their connections.

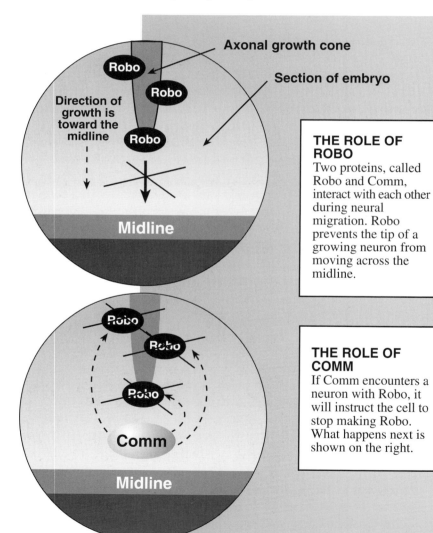

Axonal growth cone

Section of embryo

Robo

Robo

Direction of growth is toward the midline

Robo

Midline

THE ROLE OF ROBO
Two proteins, called Robo and Comm, interact with each other during neural migration. Robo prevents the tip of a growing neuron from moving across the midline.

Robo

Robo

Robo

Comm

Midline

THE ROLE OF COMM
If Comm encounters a neuron with Robo, it will instruct the cell to stop making Robo. What happens next is shown on the right.

Figure 26

WHAT A MIDLINE IS
A midline is a group of molecules and cells that literally divide an
embryo in half. It possesses numerous signals that serve to orient
migrating neurons as they create various neural structures.

Repulsion

1

Groups of neuron tips express Robo protein ("R" in the drawing) on their surfaces. The midline possesses varying concentrations of Comm ("C" in the drawing). As long as Robo is on the surface, the neuron will not migrate.

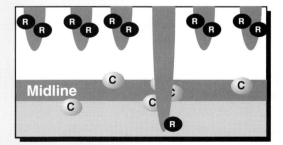

Attraction

2

As described on the previous page, if a neuron in a specific location encounters Comm, the neuron will stop making Robo.

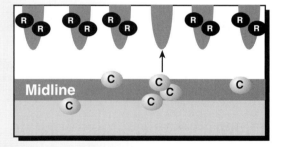

Migration

3

With the loss of Robo, the neuron tip is free to migrate across the midline. As it moves farther from Comm, however, more Robo is made. The repulsive effects are restored and the tip cannot migrate backwards across the midline.

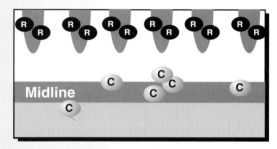

a scared rabbit, or come right along side the midline and then move down. It was not long before researchers suspected that our ability to have right- and left-hand nerves – as well as a brain – has a great deal to do with the interactions between the midline and the axon growth cones that encounter it. What are the molecular cues that guide such attractions and repulsions?

Recently, the genes and proteins that govern these finicky interactions have begun to be characterized. Much of the initial work started out with the isolation of a gene called *robo*. Scientists found this gene by creating a series of mutations in fruit flies that messed up their neural bilateral symmetry. They called the mutation *robo* for the word roundabout, so-called because normally orderly neurons criss-cross the midline in a disorderly fashion in animals with this mutation. Very soon after, researchers isolated another gene called *comm*, short for *commisureless*. In this mutation, the neurons don't ever cross the mid-line, but rather run down its sides like a parallel interstate. It turns out that the interactions between the proteins made by *robo* and *comm* reveal a great deal about how neurons get hooked-up and create their talented bilateral symmetry.

It was soon found out that the *robo* gene makes a protein that becomes tethered to the axonal growth cone and then one of those interesting sensors. To what exactly does this sensor respond? The Robo protein tells the nerve that carries it to "stay away from the midline". The more Robo protein there is on the surface of the growth cone, the stronger the repulsion cues become. This result gave researchers their first clue as to the molecular cues guiding migration patterns at the midline.

The second clue came by examining the protein encoded by the gene *comm*. It was found that Comm does not exist on any growth cone, but instead is expressed near the midline. In fact, it is expressed at different concentrations all along that midline. Exactly what is the function of *comm*? The seminal result came from examining its interactions with neurons carrying the Robo protein. If Comm binds to a neuron with Robo protein on it, the Robo neuron receives the message "get rid of any Robo you currently have and stop making any new Robo." When that occurs, the neuron obediently obeys the instruction. When it ditches its Robo protein, it loses its repulsion to the midline, and then behaves in the opposite manner, starting to migrate toward the midline, eventually crossing over, then migrating away on the other side (see Figure 26).

Here's where the results get interesting. As the axonal projection over-shoots the midline, it starts to encounter Comm protein less and less. Eventually it does not encounter Comm at all. How does it respond? Just as you might expect; it once again starts to make more Robo protein. And the more Robo is made, the more the neuron stays away from the midline. As long as there is no Comm, the result is that the neuron crosses the midline and is never allowed to look back. Voila – you have begun the neural structures that eventually result in the bilateral symmetry.

An interesting feature of this system to note is the great power of the midline and the matrix. I said that the *comm* gene is expressed in different concentrations all along the midline. Where *comm* is not expressed, there will never be an anti-repulsion signal. No neurons will cross in that space. Neurons will only ever cross the midline in those areas that have a minimal concentration of *comm*. Here then is another great example of how genes expressed in neurons and genes embedded in embryonic structures talk to each other.

Taken together, the isolation of the genes for Reeler and Scrambler and Robo and Comm have given us profound insights into the brain's hardwiring. But back to our subject, are there any data to suggest that this hardwiring actually accounts for some of the cognitive talents human beings possess? How might hardwired neural movement give us our various abilities to think, and how might that shape our personalities?

Happily, the techniques of molecular biology are beginning to give us hints as to how neural architecture relates to various kinds of thought processes. To understand some of these data, we will need to discuss some of the intimate molecular details of neural movement. As this is a bit complex, I would like to make an analogy, relating a conversation that my wife and I had when discussing art and science (something that happens in our household a lot).

Thinking and The Louvre

Before it was extensively remodeled, I used to think that The Louvre museum was an odd place to put paintings. The reason for my opinion had to do with contrast. Here was this perfectly ordinary palace housing some of the world's greatest art. Some of this art is very fluid and rhythmic and energetic; even some of the medieval carvings, reminiscent of the type Dante mentions at the start of the First Cornice, have this sense of vitality and discharge that belies the century in which they were made. But for the most part, the masterpieces used to be all stacked high in boring halls, in out of the way places or not shown at all, the ultimately contextual slap in the face.

I am certainly not alone in this opinion. When we heard the museum was going to be remodeled, my wife and I began talking about how they were going to address this contradiction. We wondered what the museum would look like if it were as fluid and as interesting as the masterpieces inside. We began to put verbs to our imaginations, and then we got silly. What if we were engineers and architects and were charged with the following task: create a skyscraper housing an art museum to reflect the strength of the art inside it in, say, a kinetic fashion. What if the building could (1) move of its own accord and (2) change shape as it moved *without interfering with the paintings or the patrons viewing them.*

Just imagining the interior of such a structure is a challenge. In terms of motion, one would need engines hooked-up to some kind of energy source, all interfacing with structural elements outside the building to allow traction. There would have to be stabilizing structures and lots of communication between the various offices and galleries, all to prepare the insides for a change in motion. There would have to be demolition experts who could continuously tear down a given structure. And one would need on-the-spot architects and construction workers to rebuild what had just been torn down into a new design. One would need recycling programs for the demolishers and builders, so that new raw materials would not have to be transported internally every time the building changed shape. Also, there would have to be a way of preserving the function of the offices and galleries while their walls, electrical systems, and plumbing were being reconfigured.

Does all this sound far-fetched? If you were an engineer asked to create such an amazing museum, you might quit the company and go find something easier to do. My wife and I soon gave up our musings too, because creating a complex building that moves and changes and stays exactly the same is an impossibility. While the word impossibility is a good adjective to use for manmade structures like art museums, the task is not an impossibility *per se.* Amazingly, most of the cells in an adult human body can perform these tasks, preserving specific function while allowing for movement and shape change. Back when you were an embryo, *all* your cells were capable of performing such neat tricks. We just talked about one such case, the migration of neurons in setting up parts of the human (and fruit fly) central nervous system. Now I would like to pursue this activity at a more structural level, directly addressing the migratory engines within neurons. We shall see that these cells really do stay fully functional, even though they are constantly moving, constantly changing shape, constantly tearing apart their insides and re-forming as they migrate throughout the body. But we mention one more gene in a seemingly tireless parade for another reason. There is a giant surprise at the end of the presentation of these data: we have finally isolated a gene intimately connected to the kind of thinking that allows The Louvre to exist, for painters to paint, architects to build, and for sculptors to carve.

To discuss all this stuff, we need to talk a little bit about internal cellular architecture and then a specific way proteins control their activity. Then we can describe data that seem to be as wonderfully contrasting as a medieval bas-relief in an old palace, and as miraculous as a museum building moving and changing like a bean bag.

Some background – architecture

One sees amazing things when looking inside the cytosol of a human cell. One is first struck by its skeleton. As you may know, a cell has an internal scaffolding composed of many different kinds of molecules. During locomotion, a great deal of destruction and construction of this skeleton goes on, all at the same time. There are enzymes that act like construction workers and there are proteins that function like steel beams and re-bars. One of the most common of these structural proteins is called actin, found in every cell of the human body, and in most eucaryotic cells (those with a nucleus) that call Earth home. It's amazing stuff, this actin. It is composed of long chains of subunits, strong enough to provide structural integrity, flexible enough to be torn down quickly. Once depolymerized, the various subunits can be reassembled into different shapes for a specific need.

Neurons use actin to maintain their static adult shapes and, during development, for migration. The actin at the leading edge – where the movement is taking place – becomes organized into a dense patchwork of molecules. This polymerization helps drive forward motion, which can take the form of spikes (termed filopodia) or sheets (termed lamellipodia). The actin is then depolymerized and reassembled into a slightly different configuration, providing motion to the protuberance. If you have trouble visualizing this, just remember the crawling motion of an ameba from your grade school biology class, the cell membrane stretching out arm-like in one direction, the cytosol pouring into the arm like sand in a bag. The tip and lateral sides of the stretched out arm are where the actin polymerization and depolymerization take place.

As we just discussed, developing neurons send out growth cones, which in concept are similar to those arm-like projections of the ameba. These growth cones are encountered whether they are working to create spinal neurons or specific regions in the brain. But the extraordinary fact is that, even though a neuron possessing a growth cone is constantly being assembled and reconfigured, its internal functions are preserved. This is analogous to the office building of the Medinas' imagination functioning normally while it is being gutted and re-formed on the inside, and moving on the outside. Scientists are deeply interested in what regulates this flexible actin scaffolding in migrating neurons.

Some background – turning proteins on and off

Processes like actin polymerization are highly regulated, as you might expect, as are the enzymes that govern the actin regulation. Before we talk about some data regarding control issues, we need to talk in general terms about how molecules – particularly proteins – turn themselves on and off. And to do that, I would like to return briefly to the world of my junior high, and probably yours as well.

Long before I knew anything about The Louvre, I was creating artwork with the specific intention of inciting behaviors in some of my third grade friends. One of my favorite activities was to draw a picture of a bull's eye in red crayon – and then write the words "kick me here" with an arrow pointing to the bull's eye. I would then apply a piece of tape to the artwork, and gently place the note on the back of one of my friends. It sometimes worked like a charm, with my friends all taking turns to hit the bull's eye on my poor unfortunate victim. More than a few times I felt an obnoxious gluteal kick to my person, discovering that someone had transplanted the incendiary note back to me.

Believe it or not, cells use the same kind of immature signaling to incite specific behaviors in many of their proteinaceous colleagues. Only they don't use cellulose and crayon, but little collections of molecules that go by the general moniker of side groups. One of the most common side groups is a phosphorous atom surrounded by oxygen molecules, a collection we call a phosphate group. When one of these side groups is "taped" to the backside of certain proteins, something generally happens to them. Another molecule might come along and "kick" the hapless protein, causing the protein's function to change, turning it on or off. The protein can respond itself to the presence of the side group as well, with no need for assistance from a molecular colleague. This alteration can happen in many ways, from changing the shape of the protein to its actual physical destruction.

The data

With these concepts of actin polymerization and side group addition in mind, we are ready to turn to the data on neural motion. As I said earlier, this exploration has yielded some rather unexpected insights into the molecular biology behind human behavior. The data consist of a group of explorations done by people in several laboratories, each focusing on different genes and proteins. Three separate genes have been isolated that, taken together, tell our interesting developmental story.

Gene no. 1

The first gene is called *cofilin*. The protein the gene encodes was found to be very important for a number of processes that were previously shown to involve actin. One of those processes is cytokinesis, that amazing process whereby a cell splits in two identical halves during replication. In the world of the test tube, Cofilin works with actin by binding both to long chains of the proteins and to individual subunits. Like putting a piece of dust into a finely tuned watch, the binding of Cofilin to actin actually stops the functions of actin. If Cofilin binds to a subunit, the actin cannot assemble into long chains. If Cofilin binds to a long chain, it causes the chain to disassemble into small subunits; other *cofilin* proteins then find the newly liberated subunits and bind to them. As you can see, actin functions soon grind to a complete halt.

Cofilin is a dangerous molecule, and, like all potentially lethal proteins, is highly regulated. It must have the ability to be turned off so that its dangerous depolymerizing activity can be regulated. And how is this inhibition achieved? Through the addition of one of those side groups we mentioned. When Cofilin receives a phosphate group, *cofilin* is inactivated. When Cofilin loses its phosphate group, the protein is turned on. Here we have one of those deconstruction processes working in the cell to destroy existing scaffolding, and it can be turned on or off. So the next question is, what gives Cofilin the rescuing phosphate in the first place?

Gene no. 2

The second gene, isolated in another laboratory, goes by the simple name of *rac*. This Rac protein has the opposite function of Cofilin. Rather than destroying existing structures, Rac has the ability to supervise the construction of actin polymers. No one knows exactly how Rac does this, though it appears to be as regulated as Cofilin and for the same obvious reasons. One suggestion was that Rac controls some process that has the ability to call the shots for Cofilin. That insight appears to be the correct one, and the description of the third gene actually sheds light on the entire system.

Gene no. 3

The third gene is one of the most interesting molecules in developmental biology. It is called LIMK-1 kinase. A kinase is a class of protein that adds phosphate groups to specific targets. LIMK-1 is no exception. The question was: what are its targets?

The answer to that question turns out to be important for neurobiologists. First of all, there is a *lot* of LIMK-1 kinase in neural cells, a fact that is always a heads-up to molecular biologists. This ubiquity led researchers to perform one of those knock-out experiments, where they looked at the behaviors of mice whose LIMK-1 had been destroyed. They found an extraordinary thing. Mice without LIMK-1 lost a cognitive ability that every artist must possess in order to create. They lost the talent of visuo-spatial cognition. What is that? Visuo-spatial cognition is the ability to integrate individual parts of a given subject into a whole, cohesive picture. You use this ability every time you try to put a jigsaw puzzle together. Artists use this in some form every time they paint, build or sculpt. There are very few genes ever isolated whose biochemical function is so concretely linked to the ability to think in a certain way. Consequently, a lot of people are studying it, including those working on actin regulation. Indeed they are the fortunate ones, for they hit paydirt in a remarkable series of coincidences. LIMK-1 kinase does indeed add phosphate groups to a specific target – researchers found that it actually adds phosphate groups to Cofilin! And when these groups are added, actin stops depolymerizing. So one important molecular regulation pathway of actin destruction has indeed been elucidated.

However, I did say the words "remarkable series of coincidences". The next question addressed the newest results: what molecule regulates

LIMK-1 kinase? Well, this question was also answered in this line of research, which makes this story so interesting. It turns out that Rac, the second protein we talked about, controls LIMK-1 kinase. Rac activates LIMK-1 kinase, causing it to place those inhibiting phosphates on Cofilin. Rac is one of the master controlling sequences in a chain of regulating molecules. When Rac is activated, Rac tells LIMK-1 kinase to "find Cofilin and tell it to quit destroying actin filaments" (if that sounds a bit confusing, see Figure 27).

What this means

If you are still with me after all this molecular discussion, you might be tempted to ask if the discovery of this little regulatory circuit is a big deal. There are hundreds of thousands of proteins in a given cell, after all, and millions of neurons trying to make connections in an embryo's developing head. Obviously LIMK-1 kinase plays a central role in the actin story, but what does that have to do with the real world of human thinking?

I stated before that a surprise awaited researchers interested in actin polymerization and brain development regarding LIMK-1. There is a condition known as William's syndrome, a multi-gene deletion disease characterized by mild retardation and a loss of visuo-spatial cognition. This is the same talent that was lost by those mice that suffered LIMK-1 kinase knock-outs. It is not difficult to see why researchers became excited about this similarity, and a number of laboratories went to work isolating the gene. They found it, and, as you may have guessed, it was the human LIMK-1 kinase gene.

This is a big deal because it gives us some of our first glimpses into how the human brain at the molecular level views its not-so-molecular external environment. People with William's still have a functioning brain; however, they have lost particular cognitive abilities, deficits linked to LIMK-1 kinase, a molecule deeply involved in neural migration.

Integrating these data

Taken together, *reeler* and *scrambler* and *robo* and *comm* give us an interesting, if convoluted, view of how genes function to hardwire specific neural regions of the brain. Don't be surprised if it all sounds confusing. Scientists don't understand much about it either. With present technology, we only

Molecules of human thought ...

The structure of actin is important in maintaining neural cell shape and function during embryonic development. Its importance in human thought processes is shown in William's syndrome, a multi-gene deletion disease characterized by mild retardation and a loss of visuo-spatial cognition. The causative agent of this disease, a mutation in the gene encoding LIMK-1 kinase, has been isolated. Shown below is how LIMK-1 kinase functions normally to regulate actin structure.

It works in concert with other proteins, including Cofilin and Rac, to accomplish its goals.

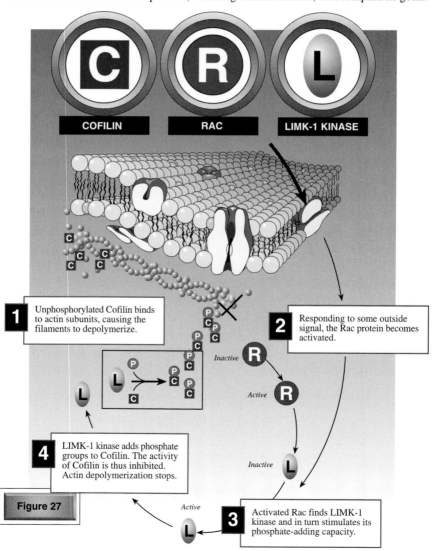

COFILIN   **RAC**   **LIMK-1 KINASE**

1 Unphosphorylated Cofilin binds to actin subunits, causing the filaments to depolymerize.

2 Responding to some outside signal, the Rac protein becomes activated.

Inactive   *Active*

4 LIMK-1 kinase adds phosphate groups to Cofilin. The activity of Cofilin is thus inhibited. Actin depolymerization stops.

Inactive

3 Activated Rac finds LIMK-1 kinase and in turn stimulates its phosphate-adding capacity.

Active

Figure 27

have the vaguest whispers and suggestions as to how migrating neurons use molecules like those mentioned above to create the most sophisticated thinking machine on Earth. From matrix to neural cells, from receptors to the molecules that bind to them, from powerful attractors to equally sturdy repulsors, there are many concepts to be understood in just these interactions. And there are thousands of other players we have not isolated. The list may seem so complicated that it is easy even for scientists to lose sight of the overall goal of these genes, namely to create a survival organ capable of living in a hostile terrestrial environment.

I can offer something of a summary, however, and in the form of an illustration. It is found in a verse from Canto XII, sung in the First Cornice of *The Purgatorio*. The subject is an interesting tale from Greek mythology, describing how many complex interactions give rise to an emergent, singular story (which in the end is what brain development is all about). Here's the verse:

Ah, mad Arachne! so I saw you there
Already half turned spider – on the shreds
Of what you wove to be your own despair

This stanza describes a vicious contest to which every Florentine could relate, a battle involving garments, fabrics and tapestries. Here stood Arachne of Lydia, tall, beautiful, and very gifted as a seamstress. "No god or goddess could best my work," she boasted to her friends, referring to the goddess Athena, a gifted weaver in her own right, "I am the greatest weaver in the world." Arachne's fame spread to the ends of the Earth, as did her pride, which was why Dante included her story in the First Cornice. So great was her reputation, and infuriating her attitude, that Athena put on a disguise and went to Arachne's house. "I challenge you before the gods to a contest," the offended goddess declared, "we shall both weave a tapestry, and we shall compare our efforts. Then we shall see who is the best in the world."

Arachne accepted the challenge, and both women soon set to work, taking myriad pieces of thread and molding them into works of high art. Arachne's tapestry showed scenes of the gods in an unfavorable light, depicting incidences of failure and disparage. Athena's fabric told the opposite story, illustrating scenes of the god's successes and power.

"What are you doing!" Athena roared, as she beheld the awful perfection of Arachne's work, and the satire emerging from her complex threads. It was not a question. "I will not allow such blasphemy with such work!" she breathed and with a flourish, the deeply offended Athena changed Arachne into a spider, even as she spun her threads. We retain vestiges of this myth in the scientific name for spiders, which as you know are called arachnids. "Of what you wove to be your own despair" is Dante's way of relating the penalty of pride for even the most justified of talents.

Whether ego problems exist or not, it is obvious that a great deal of skill is required to turn simple pieces of thread into powerful works of art, or into sticky gill nets capable of feeding an arthropod's appetite. And it is here that the analogy of tapestry applies to our discussion of brain development. Even though we have just gone over a number of complicated molecular interactions in describing these six genes, their overall goal is quite simple to understand: they work together to create a survival organ with powerful emergent properties of cognition. One can think of the neurons weaving in and out of a developing embryo as one thinks of individual threads bobbing in and out of Arachne's loom to create an emerging story: not a comment about gods, but the god-like ability to think. The patterns of neural linkages are no less solid than the patterns in the thread of a tapestry; when taken together they create the most complex story on the face of the Earth.

For the purposes of this part of the chapter, these solid patterns make up the genetic side of the nature/nurture aspects of human personality. This is part of the story of how the brain wires itself together, creating a logic board that will last a lifetime. It is a view of the nature side of human thinking, if you will, revealing a small part that our genes play in the creation of human personality.

Of course, nature isn't the whole story. As you recall, we used the analogy of babies with cataracts to show just how dependent neurological processes are on environmental experience. What makes humans so interesting is that the internal brain processes responsible for our sense of self can also be deeply affected by our environment. In fact, it is foundational to modern psychiatry that the sum total of our experiences as we go through life affect not only *how* we behave but also *who* we are. This means that our sense of self is profoundly connected to our abilities both to learn things and to recall and react to events once they have occurred. Along with our hardwired circuit board, the sum total of these memories and abilities gives us our identity. It's a survival issue; we are a species that learned to conquer a world using only a cortex and an opposable thumb.

Since we have already briefly addressed how the brain solders together its neural logic boards in the womb, we will turn to the nurture issue itself. If we really are the extreme species specialists in figuring out ways of learning from our world, how do our brains do it? We have seen that a great deal of hardwiring must occur in the womb before our brains become competent to endure the Earth's hostile environs. But after the die is cast, what allows the organ such rich flexibility that our very identities can be contoured by outer experiences? In other words, how do humans *learn* things? And how does that learning shape our sense of identity? These are our next great questions as we seek to understand the nature/nurture interactions that create human personality.

Whereas there is no firm biological answer to the latter question, there is a great deal of biological information to relate about the former. In this section, I wish to discuss the genes that allow us to bend to our environment, describing the cells and processes used in human learning. As ever, we will start with the groups of cells in the brain and end up in their neural nuclei. Along the way, we will discover that a long road still exists between the genes that mediate how we learn things and the ever-present sense of our own identity.

Is there a difference between apprehension and memory?

We start our discussion of the environmental inputs by trying to distinguish between the concept of learning and the notion of storage. Obviously, construction of an identity will involve both learning and memory, but is there a meaningful biological distinction between the two?

Most researchers have concluded that learning is part of the overall concept of memory. Learning may simply be the process of acquiring certain kinds of memories, the process that occurs when information is first presented. Consider the following example: Dante gives many illustrations of the dangers of pride as the poets tour the First Cornice, and some examples are fairly gruesome. One involved Tomyria, ancient queen of the Scythians, who thought that Cyrus, the Persian emperor, was a bloodthirsty creep. She defeated him in battle, cut off his head, threw it into an urn filled with human blood and then shouted into the urn. "Drink your fill, your Highness!" How gruesome indeed.

And perhaps memorable. You may have never heard of Tomyria before you read the preceding paragraph, in which case neurons in your brain have just acquired a new piece of information. That would be learning, no question about it, a first presentation of a strident historical personality. But what will happen when you turn the page? If you are to recall who Tomyria is later, you will have had to store her cruel visage somewhere in your brain. Memory may thus be defined as "what happens afterwards", the gradual process of massaging information into a form that will be maintained over time. This is what the researchers mean when they say the concepts of learning and memory are inextricably linked. Without learning there will be no memory. Without memory, no learning can be said to have taken place.

If even part of that paragraph stands the test of time, then trying to understand how human beings learn something also requires an understanding of how humans remember things. And that lands us on some familiar ground; every chapter in this book has had sections devoted to memory in some form or other (our exploration of the ingredients of consciousness, for example, has included the various kinds of memory systems that exist in human brains). It is high time we took a closer look at the tissues and genes involved in these systems.

As we have discussed, the multiple memory systems humans possess appear to be governed by their own distinct neural pathways. There is a great deal of disagreement amongst researchers, however, about how specific neural highways contribute to the formation of particular types of memory. Happily, as our technologies and subsequent experimental designs have improved, some of these disagreements have begun to be settled. Indeed, the view that these results afford is fascinating.

One example of these data occurred recently, and involves the relationship between working memory and long-term storage. As you recall, working memory (formerly called short-term memory) is like an ever-present neural desktop, finite in scope, useful only for the temporary storage of input. But the emphasis is on the word *temporary*. If we are to remember information for longer than a few seconds, we are forced to put it into some kind of long-term device. As you might suspect, there has to be communication between these two memory models that can be fleshed out in molecular terms. Recently, some of the pathways between these two talents have been uncovered.

Researchers examined one kind of learning, the acquisition of a motor skill, and tried to view what the brain looked like as first the person learned the skill, and second the skill was stored. Motor skills were chosen for a simple reason. It has long been known that as one practises a skill, such as playing the piano, the stiffness of the limbs decreases, movements of the arms and hands become smoother, and anticipation of what occurs becomes almost reflexive, as if the body were creating and then following some kind of internal map. Motor skills are therefore easy to study, looking at the initial learning, and then observing the changes as the person practises. Indeed, this internal map idea is what appears to be happening; the brain creates an internal model of the task dynamics, enabling the brain to predict and compensate for the mechanics of the behavior. The researchers could actually watch the brain in action using brain imaging techniques. They found that within six hours of the motor skill being learned and practised, the brain was engaging different regions within it to guide the task compared to when the skill was being first learned. Specifically, there was a shift from the prefrontal regions of the cortex, where the behavior was first introduced, to other areas deeper in the brain (such as the cerebellum). It was found that it takes about six hours for this transfer to be completed. The researchers hypothesized that this shifting occurs as the neural representations in working memory are gradually replaced by representations in the long-term storage devices. The brain imaging studies showed exactly which neurons were involved, and even gave hints as to how long it took for the transfer to be completed.

These results are extraordinary, primarily because the researchers were able to demonstrate specific changes in the areas of the brain responsible for a particular behavior. That's the Grail of course, associating discrete cognitive talents with groups of cells deep in the brain, and it portends great excitement for the future. As the technology is increasingly refined, we will better and better link models of human thinking such as "working memory" with real live tissues.

The genes of learning

But real live tissues are not a singular substance in the brain of course. Real live tissues are composed of real live neurons – packed full of real live biochemicals – and so the question of "how" also has molecular components. Other groups of researchers are tackling some of the most intimate questions that can be asked about learning: what are the relevant cells doing when the brain decides to educate itself? If parts of the human personality are shaped by the environment – that really means learning – one might reasonably hypothesize that permanent changes must occur in a given set of neural cells if the input is to be long-lasting. These are command and control issues of course, and so these questions are laid at the strategic planning center of any neuron, the nucleus. Put succinctly, what genes are involved in learning?

The powerful techniques of molecular biology are giving us answers to that question, answers that even several years ago might have seemed

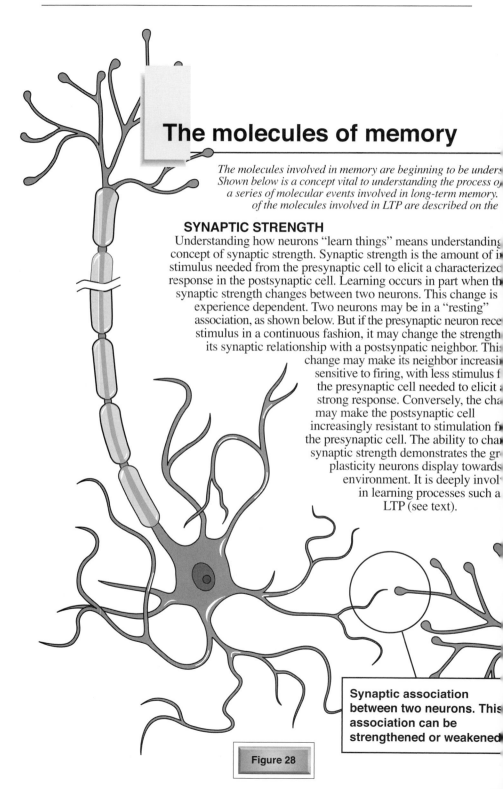

The molecules of memory

The molecules involved in memory are beginning to be under
Shown below is a concept vital to understanding the process o
a series of molecular events involved in long-term memory.
of the molecules involved in LTP are described on the

SYNAPTIC STRENGTH

Understanding how neurons "learn things" means understanding
concept of synaptic strength. Synaptic strength is the amount of i
stimulus needed from the presynaptic cell to elicit a characterized
response in the postsynaptic cell. Learning occurs in part when th
synaptic strength changes between two neurons. This change is
experience dependent. Two neurons may be in a "resting"
association, as shown below. But if the presynaptic neuron rece
stimulus in a continuous fashion, it may change the strength
its synaptic relationship with a postsynpatic neighbor. This
change may make its neighbor increasin
sensitive to firing, with less stimulus f
the presynaptic cell needed to elicit a
strong response. Conversely, the cha
may make the postsynaptic cell
increasingly resistant to stimulation f
the presynaptic cell. The ability to chan
synaptic strength demonstrates the gr
plasticity neurons display towards
environment. It is deeply invol
in learning processes such a
LTP (see text).

Synaptic association
between two neurons. This
association can be
strengthened or weakened

Figure 28

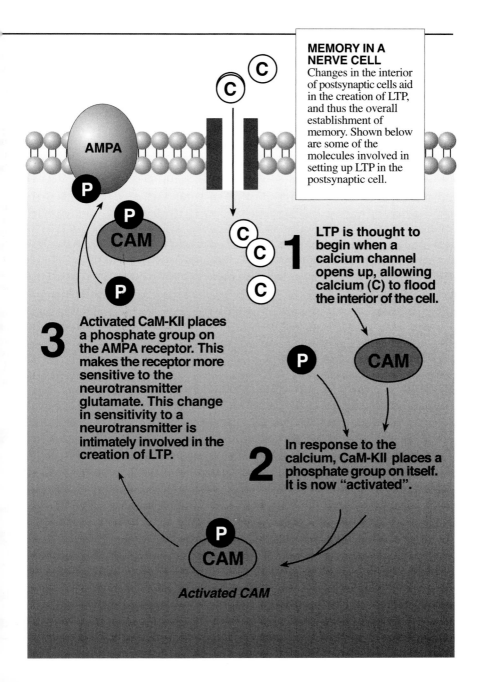

MEMORY IN A NERVE CELL
Changes in the interior of postsynaptic cells aid in the creation of LTP, and thus the overall establishment of memory. Shown below are some of the molecules involved in setting up LTP in the postsynaptic cell.

AMPA

CAM

1 LTP is thought to begin when a calcium channel opens up, allowing calcium (C) to flood the interior of the cell.

CAM

3 Activated CaM-KII places a phosphate group on the AMPA receptor. This makes the receptor more sensitive to the neurotransmitter glutamate. This change in sensitivity to a neurotransmitter is intimately involved in the creation of LTP.

2 In response to the calcium, CaM-KII places a phosphate group on itself. It is now "activated".

CAM

Activated CAM

preposterous to consider. But there is growing evidence that a number of genetic processes, strikingly conserved from one animal to the next, actually help mediate learning. And a number genes have been isolated that prove the point. We will talk about two overall processes that involve these genes, changing our level of the discussion from groups of cells to groups of individual synapses (as you recall, synapses are the spaces between neurons that are in active communication with each other). We will start with a concept known as long-term potentiation or LTP, and then consider several genes known to be intimately involved in the learning process.

The strongest will survive

As we have already discussed, there is strong evidence that neurons in the area of the brain known as the hippocampus are deeply involved in the storage of memories. One region of the hippocampus that has received a tremendous amount of attention is termed CA1. Scientists assume that many of the neurons in this region of the brain must experience long-lasting changes in order to mediate long-term memories. That assumption turns out to be correct.

A process undergone by many neurons in this region is called long-term potentiation, or LTP (see Figure 28). The process of LTP involves a persistent change in the strength of synapses between the neurons involved. To understand exactly what that last sentence means, and the genes that lie behind the idea, consider the following anecdote.

When I first met Kari, the woman who would later become my wife, I was dating someone else. And so was she. But I did not forget Kari. She is physically very beautiful, a talented Emmy-nominated composer and one of the nicest people I have ever met. When both of us found ourselves "available" six months later, I immediately asked her out. We had a great time, and I began thinking about her more and more. Turns out she was feeling the same. I asked her out again, and soon we were seeing each other regularly. It got so that every time we met my heart would pound, my stomach would do flip-flops, and sweat would appear on the back of my palms. I knew I was falling in love. Eventually, I didn't even have to see her in order to get the raise in pulse. Just a picture would do, or the smell of her perfume (Chanel number 5), or the building that housed the music studio where she practised. Eventually just a thought was enough. That was eighteen years ago, and, I have to admit, when I pick her up at the airport after she has made some trip, I still get those same flutters, sweaty palms and elevated pulses. Indeed, after all these years she has had to endure living with me, I consider myself the luckiest man in the world!

What was happening here to effect such long-lasting changes? With increased exposure to this wonderful lady, I became increasingly sensitive to her presence, my reactions made greater over time, needing steadily smaller cues to elicit stronger and stronger responses (a *building*, for heavens sake?). Moreover, the effect has been long term, having had the tenure of almost two decades. Who can understand the mystery of two people falling and staying in love, even if the person trying to organize the experience has first hand knowledge?

Leaving the whys of the hearts to the poets and the psychiatrists, I have an ulterior reason for relating this short history of my marriage. This idea, that increased exposure results in stronger reactions, lies at the heart of the process of LTP. Consider what happens when two hippocampal neural cells in synaptic association decide to learn something (remembering that a pre-synaptic cell gets a signal and then sends it to the post-synaptic cell). If the post-synaptic cell is continually stimulated by associated pre-synaptic neurons (neural cells can have thousands of these associations with one another) an extraordinary thing occurs. Just like falling in love with someone, the post-synaptic cell becomes more responsive to subsequent stimulation by the same source. Whereas at first I needed a date, after a while I only needed a photograph to send me into the dizzying whirls of romance. Neurologically, this increased sensitivity means that, over time, the pre-synaptic neurons do not have to send out very strong signals in order to elicit a strong response. A change has come over the association, with the post-synaptic cell increasingly sensitive to inputs by pre-synaptic neurons. What has happened is that the synaptic strength (that's the formal term) between the neurons has increased. This change in sensitivity can last a long time, sometimes hours, sometimes days, and to mark the fact we call the process long-term potentiation, or LTP. Many researchers believe that a memory occurs because of the persistent, even permanent, changes in the strength of synapses.

The genes of memory

What an extraordinary thing to say! It means that we are beginning to get a handle on the cellular processes that make up our memories, and, ultimately, who we are. And that means we are beginning to understand the molecules involved in the "flexible" environmentally responsive side of the human brain. Because we could observe what was happening at the cellular level, it was only a matter of time before we marched into the nuclei of those neurons involved in these associations and started looking for relevant genes.

We have found out so much. One important gene goes by the name of CaM-KII (short for, get ready for another convoluted name here, calcium-dependent protein kinase II), and an explanation of its biology helps us to clarify some important issues in trying to understand LTP. It is not the only molecular process involved in the LTP story, but it is one of the best understood. There are three players that interact with CaM-KII:

Calcium. As you might expect from its name, the atom calcium, the same atom used to make human bones, is important both for CaM-KII function and in LTP in general. There are large numbers of calcium atoms floating around not only our bones but also our neurons. These atoms are deliberately kept out of the neurons by molecular "borderguards", proteins that sit on the surface of a neuron, watching specifically for calcium, making sure the atoms stay out.

AMPA receptor. This is an interesting protein that sits on the post-synaptic cell membrane. You are probably used to the word receptor by now, a molecule whose function is normally initiated by having another molecule (called a ligand) bind to it. AMPA is no exception. AMPA is thought to bind to a ligand called glutamate (in fact AMPA is a subtype member of a class of proteins called glutamate receptors). However, the AMPA receptor is a bit finicky, which means it may or may not bind to glutamate, depending on the biochemical mood it's in. We'll discuss what makes it happy in a minute.

Phosphate groups. Remember these? You might recall from our discussion of Cofilin that these useful groups act like those "kick me" signs I used to put on my poor friends in junior high school. These side groups are useful for more than just working with actin filaments. If a phosphate group binds to an AMPA receptor, the receptor's function is altered. A phosphate can also be added to CaM-KII, and its function is altered as well.

How it all works together

With CaM-KII, calcium, AMPA, and phosphates in mind, we are ready to discuss how LTP works, and learn how neurons learn things (see Figure 28 for a complete description). Most researchers agree that LTP begins when the postsynaptic cell gets "hungry" for calcium and starts to drink it in. This occurs because those borderguard-like proteins on the surface of the neuron turn themselves into a specialized channel (they actually change shape), creating a hole in the membrane. The calcium comes rushing through that opening.

This onslaught of atoms into the interior is felt throughout the entire cell, including CaM-KII. In response, the CaM-KII grabs one of those phosphate groups and glues it to itself, a process termed, appropriately enough, "autophosphorylation". This one act dramatically changes the activity of CaM-KII. It now deliberately seeks out AMPA, that glutamate receptor I just mentioned. When CaM-KII finds a receptor, it grabs another phosphate group and, instead of autophosphorylating, places the side group onto the AMPA.

Now it is the AMPA's turn to become excited. As a result of the addition of phosphate, its function is altered. Specifically, the change enhances the responses of these receptors to glutamate. The system is becoming more sensitive, if you will, as if it had fallen in love and could now be stimulated by just a glance at a photo. The addition of phosphate to AMPA lasts only about an hour, but the addition of phosphate to CaM-KII lasts much longer. Part of learning, in other words, involves the specific addition of a side group to an important receptor.

This is exciting stuff. LTP has associative properties (I smell Chanel number 5, therefore I go crazy) that appear to match the cognitive process of learning as we know it. You already know that the hippocampus is deeply involved in long-term storage. But these results represent important steps in fleshing out the facets of our brains that respond profoundly to environmental inputs – the hardwired facet of flexibility that is the opposite of hardwiring, the malleable side that helps form our personalities.

And that's not the whole story

The ability of the external environment to strengthen synaptic associations points to an enormously important aspect of the biology of thought: that neural connections are *plastic*. While we have only discussed one component of the complex process of LTP, the experience dependence of the reactions is both obvious and astounding. Perhaps we shouldn't be too surprised. Neural plasticity allows a creature to learn from its environment and remember what it has experienced. Given the relative environmental

Genes and learning

The genes that mediate learning are surprisingly conserved throughout the animal kingdom. Here is how one well-characterized system works.

The genes involved in learning are being uncovered. One important sequence is called CREB (short for cyclic adenosine monophosphate response-element binding protein). CREB acts as an "alarm clock" turning on other genes involved in memory formation. It interacts with other molecules, including a complex called protein kinase A or PKA. How these molecules are configured is shown below, and how they interact in memory is illustrated on the next page.

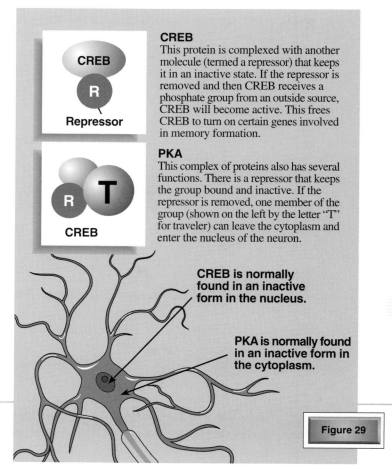

CREB
This protein is complexed with another molecule (termed a repressor) that keeps it in an inactive state. If the repressor is removed and then CREB receives a phosphate group from an outside source, CREB will become active. This frees CREB to turn on certain genes involved in memory formation.

PKA
This complex of proteins also has several functions. There is a repressor that keeps the group bound and inactive. If the repressor is removed, one member of the group (shown on the left by the letter "T" for traveler) can leave the cytoplasm and enter the nucleus of the neuron.

CREB is normally found in an inactive form in the nucleus.

PKA is normally found in an inactive form in the cytoplasm.

Figure 29

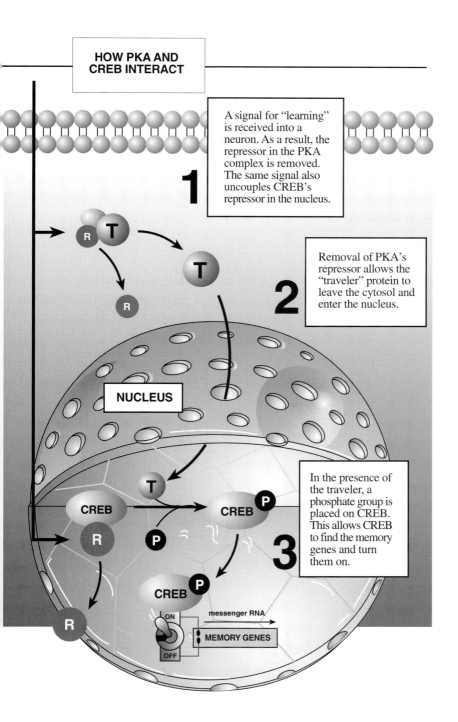

HOW PKA AND CREB INTERACT

1 A signal for "learning" is received into a neuron. As a result, the repressor in the PKA complex is removed. The same signal also uncouples CREB's repressor in the nucleus.

2 Removal of PKA's repressor allows the "traveler" protein to leave the cytosol and enter the nucleus.

NUCLEUS

3 In the presence of the traveler, a phosphate group is placed on CREB. This allows CREB to find the memory genes and turn them on.

CREB

messenger RNA

ON

OFF

MEMORY GENES

instability of most of the planet, such a talent is of obvious selective advantage.

Perhaps the most amazing thing about this plasticity lies not only in the fact that creatures learn things, but also that they tend to use the same types of genes and molecules to accomplish their educational goals. You might be surprised to know that much of what we first discovered about neural learning (and are still discovering) came from examining sea slugs. As the techniques of molecular biology came to bear on these fields, we began to isolate some of the genetic mechanisms involved in the process of making memories. As we examined the structure of genes in various animals, we uncovered an almost ridiculous conservation of sequence and function. We are going to look at one of these genetic mechanisms, describing a gene so deeply ingrained in the evolution of animal life that it mediates memory functions in virtually every organism in which it has been discovered.

Into the neural nucleus

We are going to examine a gene called CREB (to see what CREB stands for, see Figure 29) and its associated genetic mechanisms. Like those for LTP, these gene sequences are deeply involved in the creation of long-term storage. And also like LTP, the mechanism was first uncovered in a sea slug.

When a neural connection becomes reconfigured to acquire new data, a number of "sleeping" gene sequences must be activated as the neuron permanently changes its function. While this notion seems obvious enough, it really wasn't until CREB was isolated that researchers found genetic evidence for the process. It turns out that the CREB gene product acts a lot like an alarm clock. Once the CREB gene is activated, the resulting protein goes back into the nucleus, binds to a number of memory-relevant genes and says "wake-up and start making proteins". This might seem a little confusing, but remember that when a gene is activated, an mRNA is made, which first escapes from the nucleus and then finds the protein-manufacturing depots in the cytoplasm. Once there, the mRNA is decoded and the protein is made. The protein is then free to exercise its particular role. In the case of CREB, its role is to return the nucleus, find a number of inactive memory-relevant genes, and turn them on. It is truly an alarm clock, and the end result is that the nucleus is informed that a memory is supposed to be made.

The interesting thing about CREB has to do with its regulation. In neural cells, CREB can actually be found in the nucleus, but in a completely inactive form. It is rendered inactive because it is bound to another protein called a repressor. As long as the repressor remains firmly attached, CREB can stay in the nucleus all it wants, but no memories will be made. The repressor must be removed before learning can take place.

Recently, the mechanism for removing the repressor appears to have been uncovered. To understand how, we need to talk about one more series of proteins. There is a group of multi-functional proteins bound together in the cytosol called protein kinase A or PKA. The reason I say multi-functional is that, like a rock band, the group possesses subunit proteins with diverse kinds of talents. There is a subunit, for example, that can travel to the nucleus if the proper signal is given. There is another subunit which acts like one of those repressors I just talked about. This subunit keeps the group together and completely dysfunctional (including keeping that potential nucleus-traveler in the cytosol and out of the nucleus).

When it is time for a neuron to learn something, two important sets of signals are sent inside the neuron. One signal goes directly to the nucleus and uncouples the repressor from CREB. The other signal is sent to PKA and says "untether the traveler and let it do its job". The repressor relents and the traveler rushes off to the nucleus. And here's where it gets interesting. The traveler finds the now-untethered CREB and puts a phosphate group (sound familiar?) on it. With that action, CREB now finds the memory genes and activates them. Interestingly enough, one of the genes CREB activates creates an extraordinarily vicious protein. You remember that other repressor we talked about, the one on PKA that keeps the traveler from migrating to the nucleus? Once made, this vicious protein hunts down every PKA it can find and destroys the repressor. As a result, more travelers are released, and more genes are activated.

There are a couple of interesting things to note about this seemingly confusing system. Repressor proteins are useful because they make the system work fast, something you would need if you had to learn something in a hurry. To activate the memory genes, you don't have to spend a great deal of time manufacturing new CREB. The CREB is already there, albeit in an inactive state. All you have to do is take away the repressor to get the effect, and the net result is that the cell saves time. The other interesting thing is that neurons don't make a whole lot of CREB to begin with. As a result, the overall amount of things that can be encoded into our long-term memory is restricted. It is possible that such stingy controls on CREB concentrations explain the following fact. It has been known for quite a while that long periods of concentrated study are not as efficient as many short periods of learning (in terms of the number of things learned per unit time). If CREB plays as big a role in human memory as it does in sea slugs, this fact may be due to the limited amount of CREB in any one cell. The neurons may simply need more time to regenerate CREB as the cells become recruited to store pieces of information, and long periods of study simply exhaust the supply.

What all this means

Taken together, we have sequences that form both the neural architecture of the brain – *robo*, *comm*, *LIMK-1*, etc., etc. At the same time, we have sequences involved in making that architecture responsive to the environment, such as the AMPA and CREB molecules. There are many other processes, that we have not mentioned, involved in the ability of humans to learn things. Excluded from this discussion is the extraordinary ability that neurons have to increase their numbers of associations (called dendritic outgrowths) as a direct function of enriched learning environments. This interesting property is a conceptual mixture of the formation of neural architecture and environmental plasticity (see Figure 30).

We haven't talked much about the main subject of this chapter: how these sequences work together to form a sense of identity. We are getting tantalizing hints that LIMK-1 helps us to become better artists and that CREB may show us how we should study for finals. We haven't mentioned the important association between genes and identity simply because nobody knows exactly what it is. We do know that linkages exist, mostly from observations involving either accidents of humanity or accidents of nature. But the process is so complicated and the gap so wide, that at present one can only draw basic conclusions and guess at the rest. Before we move on to our discussions of mind/brain and consciousness, I would like to discuss two examples illustrating how we know that cells and identities have specific, though undefinable, linkages. Then I would like to talk about a modern "guess" concerning the nature of personality and the many neurons that cause it to happen.

The famous case of Phineas Gage

Every student who has taken introductory courses in human neurobiology has heard of Phineas Gage. He was a nineteenth century dynamite worker,

The great plasticity of the brain

The brain has such a powerful ability to respond to environmental stimuli that proper exposure can actually change its internal physical architecture. Experiments on both animals and humans prove that enriched learning environments can alter a number of important characteristics, including dendrite outgrowth. This and a few other changes are illustrated below.

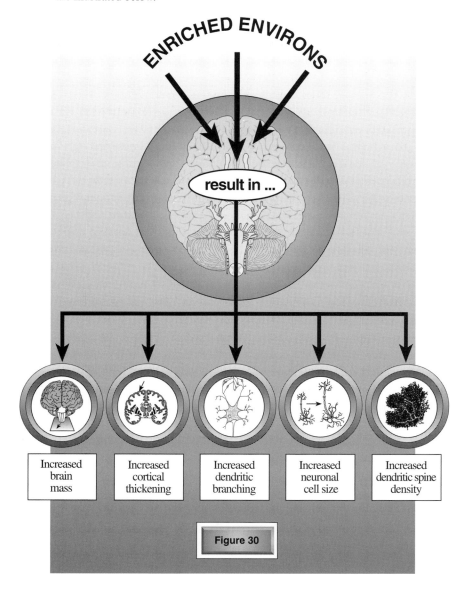

ENRICHED ENVIRONS

result in ...

| Increased brain mass | Increased cortical thickening | Increased dendritic branching | Increased neuronal cell size | Increased dendritic spine density |

Figure 30

a pleasant, hardworking person with a keen intellect and a good sense of humor. Gage makes it into modern-day undergraduate textbooks because of a horrific and bizarre accident. While at work, a freak explosion drove an iron bar through his head, affecting the frontal lobes of his brain. There are two extraordinary things to note about this accident. First, the injury didn't kill him. Second, it markedly changed his personality. After the accident, Gage began to behave in embarrassingly childlike and socially inappropriate ways. He could not follow even simple tasks when asked to perform them. He became profane and enormously impulsive, almost as if the accident had completely changed his personality. Indeed, a researcher studying him at the time concluded that the balance between his human faculties and his "animal passions" had been disrupted as a result of the accident.

Here's the second story, which relates not an accident of humans, but an accident of nature. It concerns a phenomenon experienced by many victims of strokes, called prosopagnosia, the inability to recognize faces.

The story begins with a woman rushing in to the emergency room, a horrified look on her face. She told the doctor that when she awoke that morning, strange people had invaded her house, a man and three children to be exact. The physician immediately suspected prosopagnosia, probably caused by a stroke, and ordered some tests. He eventually met the "strangers" who were in fact her husband of fifteen years and their three children. The woman had indeed suffered a stroke, and steadfastly denied ever knowing the people in the doctor's office. Unlike most of these cases, this one ended sadly. The couple eventually filed for divorce, the husband and children irreversibly estranged, the woman never resuming those parts of her identity as wife and mother.

What do both of these accidents say to us? They illustrate only the obvious: that personality and identity have deep neurological correlates. Mr Gage had a particular personality that underwent some conspicuous modifications as a result of an injury to brain neurons. The married woman with the stroke experienced a partial loss of identity, also because of an injury to brain neurons. *If you damage specific brain cells, you can profoundly alter personality and identity.*

That last sentence is an extraordinary thing to say, and for two reasons. First, concerning the distance between gene and temperament, that's about all we really know and, second, it is a big step to know even that. If you recall the convoluted road of mind/brain arguments, putting the issue of self squarely on the delicate shoulders of nerve cells is not a trivial thing to do.

There are many other examples besides dynamite accidents and prosopagnosia that illustrate the neurological roots of identity. Most people come into this world with a large palette of potential emotions and behaviors. That's what neurological hardwiring is all about. But the way the emotions are experienced, and the memories whose sum total gives us our reactions and our individual personal characteristics are based on environmental exposure. Our identities come about as a massive sum of the minute-by-minute interactions of neurons with the outer world. In many ways, it is just like our discussion of the necessity of a child's eyes to see light in order to have vision. The interaction of genes with the environment gives us the experience.

Minds and brains and Kim

The power of neurons in shaping the human being gives us one last issue to describe in our brief history of the mind/brain debate. You might recall from the last chapter that we discussed the contributions of Donald O. Hebb, who provided us with a valuable framework for fleshing out cognitive principles in terms of neurons. You might also recall that I gave a series of lectures on Hebb, which one student, whose name was Kim, took quite seriously. After an initial reaction to understanding the contribution of brains to thinking, she left my office, with a threat to come back after the next day's lecture. I will begin our last installment in discussing minds and brains by telling you she made good on her threat.

"So I want to know something," Kim declared, once again carrying her clarinet case with her. "We were talking about marching bands," she continued, "and you told me that when I think of music I am just activating groups of cells. And I told you I wasn't sure, especially about this ghost in the machine stuff."

"It was Hebb's idea to talk about cell assemblies," I related. "This concept, especially the way we consider them, is almost like an emergent property of the various multiple assemblies all talking to each other." I cleared my throat, hoping my pearls of wisdom would sink in. Suddenly she jumped up. "Yes!" she almost shouted "That's what I want to know! Doesn't an *emergent* property imply some kind of ghost in the machine after all? It's like there's a mind in there, or something. Maybe there's a marching band in our heads – maybe playing a piece of music, right? Even though you need the band to make the music, the band *isn't* the music. So the music is something other than the band."

It was a good insight. And as before I had no answer, or even a comment. We went round and round in this discussion, but the core idea was plain: by contemplating that the whole might be greater than the sum of the parts, Kim had stumbled upon the modern form of a concept called emergentism. This idea that the mind might be an emergent property of brain neurons is both new and old, and I told Kim before she left my office how cutting edge her comments would appear in some circles. The idea of emergentism is the last stop on our brief tour through the mind/brain debate.

The genes and cells we have been talking about in this last chapter provide a contrasting background for the definition of emergentism. Emergentism says "nonsense" to the strict reductionism which the isolation of these genes implies. It states that as evolution waved its patient hand across our cerebral motherboard, our circuitry combined into ever more complex entities. Eventually the entities became so elaborate that collective properties – emergent properties, if you will – came into being. That is, the property of "mind" is more than the sum of its parts. It is like Kim's understanding that music is something other than the sum total of the instruments being played. To quote Daniel Robinson, a noted physiological psychologist:

> A totally nonideological science would have to stand up and say, "We've brought the most exquisite techniques to bear on the organization and functioning of the human nervous system. And we're obliged to report to you that the richest psychological dimensions of human life are not explicable in terms of the biochemistry and physiology as we know them to date."
>
> *(Hooper and Teresi, 1992)*

In many ways, emergentism is a reaction to the reductionist philosophies that have been eating away at the soul-side of the mind/brain debate for centuries. In its modern form, emergentism is a specific response to the Hebbian form of neuron-learning, which often uses a computer as a metaphor for explaining mindness. As you might expect, the notion of emergentism is not held in universal esteem by all neural researchers. You can almost hear Kraepelin groaning from his grave in disappointment. Or is it the sound of laughter, coming from the tomb of Descartes?

Consciousness and emergentism

If you are an astute reader, you may have already thought about consciousness when you read that quote from Daniel Robinson. Indeed, some researchers have sought to invoke the presence of an emergent property to explain the fact that a definition of consciousness is so hard to come by. In one way, of course, invoking the whole-is-greater-than-the-sum-of-its-parts is a convenient way of avoiding the question. Yet people who are in the emergentism camp see consciousness as the highest achievement of brain evolution and the greatest example of a collective property. That does not mean there are no strong biological roots here. Indeed, emergentism states flatly that the property arises precisely *because* of interactions with complex biochemical systems. Perhaps we use this forced approach to define consciousness, by identifying key ingredients, simply because we have no obvious metaphor that would help us otherwise explain it. An emergentist might argue that this collective more-than-the-sum-of-its-parts property is the reason.

With these thoughts in mind, we are ready to tackle our last ingredient in our attempts to understand the notion of consciousness. In many ways, this final ingredient fits in well with what emergentists might say is the "you see, I told you so" evidence of the property. This is because this final ingredient does not concern an area of the brain, but an *interaction* between the brain and the rest of the body. Indeed, the sum total of the brain and the interactions with the body make up the ideas of consciousness and identity as a whole. For the sake of completeness, we cannot leave our discussion of these ingredients without including it.

Some researchers believe that the feedback received by the brain from the input the body experiences is a very important component of our awareness. Historically, this belief wasn't always held very highly, and the notion is hotly debated even today. William James, for example, said that when the brain felt something, the body reacted and the resulting bodily changes were felt by the brain and comprehended in a *conscious* manner. Thus, as an integral part of a feedback loop, input from the body was important in the formation of a conscious experience.

Why was that notion controversial? Here's one example. After James made his point, certain medical researchers starting asking critical questions about people they encountered in their practices. What about the consciousness of those patients who have suffered spinal lesions, and presumably have very little input coming from the body? These researchers could not believe that these people lost consciousness simply because they'd lost somatic input. Moreover, many researchers felt that the body just can't provide enough variety of signals to account for the many different kinds of emotions that humans experience. With these and other objections, the idea of a feedback system fell into disfavor.

Not permanently, however. Recently, the notion that the body plays a role in the formation of emotions and consciousness has come back into vogue. Many researchers use the arguments of natural selection as a way of ushering the interactive body-to-brain idea back into favor. They state that emotions, like consciousness, probably evolved as a way of comparing reactions in the body to the demands of the environment. They respond to spinal injury arguments in three ways. First, since very few injuries sever the cord completely, bodily throughput is experienced even in quadriplegics. Second, there are other ways to communicate reactions to the brain that do not involve neurons (through the bloodstream, for example) or that involve cranial neurons that bypass the spinal column directly. Third, memory stores in the brain can affect consciousness "as if" the brain were receiving throughput right from the body. And that's where identity issues, such as the storage devices we have talked about in this chapter as well as our ability to learn, come to the front. By activating these memories, our ability to know who we are is stabilized even if the body isn't communicating. Moreover, as more emotion systems were discovered, their possible interactions were found to generate enough varieties of feedback to account for all kinds of emotions. Some researchers have even gone so far as to say that emotional responses can alter those somatosensory representations (the homunculus we talked about previously) in the brain directly. These researchers believe that this feedback mechanism is so important that it not only plays a role in human consciousness, but in the construction of the emotions themselves.

The truth is out there

Whether one discusses how the body talks to the brain, or the brain to the body, it is clear that the interaction is so important that it cannot be neglected in our quest to define the ingredients of consciousness. There are many gaps to be filled of course, but that is true of everything in science, especially in the field of linking genes to behaviors. Without attempting to understand what ego might be, we have attempted to focus on its underpinnings, the nature of identity itself.

Taken together, it is easy to see how a discussion of pride can turn into a discussion of self. In turn, it is just as easy to observe that some of the contents of identity, which must include the ability to learn and remember, are amenable to scientific observation. By looking at the genes involved in these behaviors, we come very close to understanding why the reductionists are so enthusiastic that they can explain minds and brains in physical terms. By understanding the emergent talents such genes give us, the complexities of human personality for example, we can understand why emergentists so easily clash with their reductionist colleagues.

As the gene work illustrates, it is important for all sides to remember that they work from incomplete knowledge, no matter how strident their beliefs in the current data. Whether we are looking at brain rewiring, learning, or bodily feedback in awareness, the phenomenon of human behavior staggers under the crushing rock of mystery. That's not a bad way to look at human identity, which in the end must include the capacity for pride. Indeed, it's not a bad way to look at most of the behaviors mentioned in this book.

Conclusion

". . . I have filled all of the pages planned
for this, my second, canticle, and Art
pulls at its iron bit with iron hand.

I came back from those holiest waters new,
remade, reborn, like a sun-wakened tree
that spreads new foliage to the Spring dew

in sweetest freshness, healed of Winter's scars;
perfect, pure, and ready for the Stars."

-Canto XXXIII, The Purgatorio

"You are just like your *son!*", I heard my mother yell at the top of her voice, and for about the fourteenth time that month. She was addressing my father again, fully convinced his antics would get our family snarled into legal trouble. I was only five at the time, and she was making her comment several weeks after the "fishy hose" incident, as it came to be called, the event whereby our neighbor's station wagon turned into a short-lived aquarium.

"I'll take that as a compliment," my father grunted, dusting himself off from the floor, both of us peering out a window at our backyard. We weren't paying much attention to anything except the backyard fence, scanning intently for a dog named King. A tripped over garbage can gave away our curiosity, and the reason for our intensity.

King was a big German Shepherd, owned by our backyard neighbors, a dog who had developed a taste for the contents of our metal garbage can. Every Saturday morning at precisely 8:00 am, King came bounding over the fence like some 1960s version of a cruise missile. He went straight to our garbage can, tipped it over, making both noise and mess, and ate to his heart's delight. Figuring that the owner naturally slept in on a Saturday, leaving the poor canine ravenous without his 8:00am feeding, my father concocted a plan. He rigged up the garbage can to a battery pack, electrifying the metal; engineering it in such a way that, on physical contact, the dog's wet nose would get the shock of its life. The idea, premised on Skinner's behavioral modification protocols (so my Dad said, trying to turn this into a "learning moment"), might prevent future Saturday morning interruptions.

The can was electrified with tender loving care on Friday night, and my father got me up at 7:45 am next morning to watch the coming doggie barbecue. Eight o'clock rolled by and our excitement grew, Dad gently knocking my ribs with his elbow. Five minutes passed and no dog; 8:15, and still no dog. Dad began to get concerned because King's visit usually ran like a Swiss watch. Eight thirty and no dog, and now I was restless. Mom got up, sleepily puzzled at the two erstwhile ghouls huddled by the window, and went to make breakfast in the kitchen.

It wasn't until 8:45 am that the dog finally showed up, lazily jumping over the fence, and we immediately noted a problem. He didn't make a beeline to the garbage can as was his wont, and I heard my father curse and then say something about the owner probably feeding him. But then my dad's eyes grew wide. The dog wasn't interested in eating, but he wasn't inactive. This morning, King was interested in marking his territory in the classic urine-graffiti style of the canine world. He went to the fence post and marked his territory. He went to the side of the house and marked his territory. And then he went straight to the garbage can, a grim look on his doggie face and my dad suddenly got this knowing realization on his. "This is gonna be wild!" he whispered to me excitedly. The dog lifted his leg to urinate on the can, and, well, you don't have to know the concentration of electrolytes in mammalian urine to know that, when his urine made contact, the dog completed a mighty circuit! The dog howled – as only German Shepherds can – and ran back over the fence, his cranial neurons ablaze, his reproductive future in serious question. My dad subsequently fell off the couch we had propped against the window, laughing as hard as I have ever seen him. And my mom, who had watched the whole thing from the kitchen, came rushing to the window where we were. Probably remembering the aquarium incident, that's when she made her comment about him being just like me.

Many years have passed since I helped a friend fill up his dad's station wagon with water, and since my dad gave King his electrical neutering. I've been doing a fair amount of reminiscing about these events since the birth of our first child, Joshua, and now I wonder how many times my wife and I will make a comment like my mother's. As more great strides are taken in understanding the genes behind our behaviors, I wonder how many scientists will start saying similar things in the papers they publish.

It wasn't always like this, of course. There was a time when scientists told parents they were like a team of artists, their offspring a blank canvas, an unformed piece of marble, an unpainted wood panel. They were also told that their job was to draw a self-portrait in their children, to chisel out improper pieces, to apply the right amount of behavioral paint in the upbringing of their young. As is clear from even the small number of gene sequences described in this book, the nurture side of the question is only one component in the behavioral explanation of a human being. And it may not even be the most important piece.

How can we summarize all of this?

As I watch the behavioral complexity of even the most behaviorally inexperienced of our species, like our Joshua, I am amazed that anyone can figure out the roots of human emotional behavior. As you know, we have been exploring one brave attempt, Joseph LeDoux's model, which states that subjective emotional feelings have two components. First, an emotion system of the brain is activated. Second, we become aware of the activity. It is an interaction between specific groups of neurons and the consciousness of the person in whom those groups have become stimulated.

In each chapter of this book, we have explored the specialized emotion system of one of the Seven Deadly Sins. Besides examining the genes under-girding specific behaviors, each of the deadly sins has revealed something about the nature of the emotions under study. For example, in Chapter Two, about lust, we learned that some components of sexual arousal bypass consciousness altogether. That led to the startling conclusion that certain emotions, maybe most of them, are things that happen *to* us, rather than things we *will* into existence. It also means that we are not always aware when the emotion system has been activated. Such ignorance means that in the end we may not even know what lust is.

Chapter Three, on gluttony, revealed another interesting aspect of emotions, one that gave us great keys to understanding our own behaviors, and inadvertently, perhaps, the potential for much confusion. Emotion systems may share similar structure/function characteristics throughout the animal kingdom. Since every creature on the planet has to solve similar problems in their attempt to survive, there's a certain logic to saying that neural functions are evolutionarily conserved. And while that gives us great justification for doing experiments on animals in an attempt to under-stand ourselves, it also comes with a great warning. Animals are not people, and one of the greatest differences between us lies in the organ we are studying, our brain. While animal research may provide important clues as we seek to understand our own behaviors, it also gives us our sternest warnings against overinterpretation.

We encountered another warning about the characteristics of emotion in Chapter Four, about avarice. This concerned misinterpretation rather than overinterpretation. This is because a gene for greed has never been found, but only genes for greed's underpinnings (such as fear, the subject of Chapter Four). This led us to the conclusion that we have to be very care-ful about how we label things. A subjective category constructed by humans does not always reflect a biological function constructed by nature. This is true even for feelings; the emotion that we experience may simply be a by-product of an underlying survival process that has very little to do with how we humans have categorized something.

This idea that one has to be careful with human categorizations has a logical consequence, one that we explored in Chapter Five, when the sub-ject was sloth. We discovered that a psychological phenomenon can only be scientifically assessed if the phenomenon is shown to exist in the brain.

There is of course no question that human beings can feel subjectively tired. But these feelings may be just a reaction to input coming from another area of the brain, the area where those awake/asleep cycles of the circadian biorhythms are generated in the first place. Thus, the proper way for science to explore feelings like tiredness may begin in the pacemaker area, and then extend outward to how that input is perceived. Sloth thus has three components, the circadian rhythm area, an area where the circadian output is perceived, and the connection between the two. That's what I mean when I say that the proper level of study is the place where the phenomenon is shown to exist.

Chapter Six discussed the genetic origins of human wrath and aggression, and gave us a practical lesson on both the strengths and pitfalls of the ideas presented in this book. We found that if the emotional feelings and responses are caused by activating certain common neural systems, then we can use the emotional data to investigate that underlying system in a scientific fashion. When we experience wrath, then, we may use it to look for regions in the brain. As long as we keep our labels straight, attempting to understand human neurons rather than human categories, we even get a bonus. We obtain permission to scientifically study the link between a conscious experience and the activated emotional subsystem. Since the brain systems that make those emotional responses appear to be very similar in both animals and people, we can even use the animal data discussed in the previous chapters and apply them to humans. This can be fraught with danger, of course, as discussed in the topic of testosterone in animals and wrath in humans. This chapter serves as something of a summary for both the assumptions and warnings we discovered about the study of emotions and behaviors.

Chapter Seven underscored an important point in our desire to understand emotions such as lust, gluttony, avarice, sloth, and wrath. One of the goals of this book is to show that the brain is a powerful organ directed at a single purpose: human survival. We used the example of depression in our discussion of envy to illustrate that sometimes this purpose can be obscured by the very emotions we are seeking to understand. At this point in our psychological understanding, envy appears to be a collective reaction to one or more underlying mechanisms. One of these mechanisms has its roots in depression. Surprisingly, depression itself appears to be nothing less than a survival process worn down by chronic over-exposure to a perceived threat. That's an almost counterintuitive idea – an overheated *survival* mechanism? – considering that so many suicides come from experiencing depressive illness. But that is precisely the point. Emotions can sometimes be a distraction to the scientific study of the underlying mechanism that generates them. And if there are layers of undergirding mechanisms, as appears true with envy, these distractions can easily turn into obscuring detours.

Whether we are taking a detoured route, or find ourselves directly atop a relevant neuron, all of the previous chapters nearly ignored the overall context of human emotion and behavior – their residency inside a human personality. In Chapter Eight, we attempted to address this deficiency, using pride to talk in genetic terms about what some researchers think personality is. Critical parts of the self come from our ability to remember and react to the historical records of our lives. Therefore, understanding how brains embed information is crucial to this task. We focused on one important aspect of personality formation, the molecular biology behind the ability to absorb and encode knowledge. And with that description we ran into a familiar theme. Even a cursory review of these data shows that concepts such as pride and self and personality are truly incompletely described phenomena. Simply put, the overall framework for contextualizing emotions and behaviors is as inadequately described as the individual behaviors themselves. I am beginning to sound like a broken record, but we still have a long, long way to go before we understand the pathway between a behavior and a nucleotide.

Sketchy as they are, these efforts seek to describe the first great component of LeDoux's model, the environmentally interactive emotion subsystems of the human brain. The second great component of a subjective emotional feeling, as you recall, was a phenomenon upon whose door these emotion subsystems knocked hard. The phenomenon is human consciousness, the idea of awareness, a concept that has frustrated just about everybody who has attempted to describe it. And that includes me.

I took the most cowardly route I could in attempting to explain this part of LeDoux's model, and only in reaction to something I feel deeply. From a strict reductionist point of view, I believe that nobody knows what consciousness is. *Nobody.* There are many possible reasons for this lack. It could be that, after all these years, there is no adequate metaphor to describe such a complex phenomenon. The lack might exist because of any of the pitfalls we have discussed here, from misidentifying certain subsystems to the fact that consciousness may only be a human idea, and not a neurological one.

I did not abandon the notion entirely, critical as it is to our discussion of behavior. Instead, I attempted to explain specific neurological ingredients that must be present if consciousness as a neural concept exists at all.

Taken chapter by chapter, different ingredients make specific contributions. These contributions are listed below:

1. The contribution of short-term buffers (Chapter Three).
2. The contribution of long-term memory (Chapter Two).
3. The contribution of working memory (Chapter Four).
4. The presence of other arousal systems, and the lack of a "continuum
   of consciousness" (Chapter Five).
5. The contribution of underlying unconscious stimulation (Chapter Six).
6. The notion that different types of consciousness exist (the difference
   between conscious emotional feelings and conscious thoughts, for
   example) (Chapter Seven).
7. The importance of body feedback, and the presence of "as if"
   phenomena (Chapter Eight).

The areas in the brain responsible for several of these important ingredients have been isolated, and so, at least at a cursory level, consciousness
appears to have aspects that can be studied responsibly. The great techni-
cal problem lies in finding out how they cooperate with each other to
create consciousness, assuming that such cross-talk exists. Much controversy lies behind that last sentence, and the fact that we are still in the
middle of fighting about such fundamentals illustrates how much we do
not know.

There are many things relevant to a discussion of human behavior that
we did not mention in this book. Free will, for example, and the power of
choice in the midst of nature/nurture come naturally to mind as we seek
to overview the great distance between attitude and allele. There is a delib-
eration in staying cursory, however, and the point was not to be thorough,
as mentioned in the Introduction. The point was to show the vastness of
the ground.

I don't mean to be discouraging here, and hopefully you have found
that this tome focuses on as many genes as gaps. We have come a long way
in our perspectives since the time of the Greeks; in fact, I included a brief
history of the mind/brain debate precisely to underscore such progress.

The message in the music

Whether the mind is the brain or simply something the brain does, there
is always a great beauty in beholding a good one at work on some

important project. One of the most brilliant minds – or is it brains? – my wife and I have ever encountered belongs to a musician/composer friend of ours. We have all had a good laugh, for she too is working with part of *The Divine Comedy* in her creative efforts. Specifically, she is composing a symphonic tone-poem based on the book right before *The Purgatorio* and called, like this book, *The Inferno*. Our son Joshua is fifteen months old as of this writing, and, though he won't remember it, he has recently met our friend and heard her outline her musical idea. Joshua has red hair right now, just as I had when I was his age, and as we introduce him he starts raising his hand, seeking to be noticed by her, going through one of those interminable mysterious behavior phases common to all children. He even gets a look about him my mom recognized in me when I was Joshua's age. Mom called it the "incandescent imp" look, a look I got when my cerebral wheels were turning, a half-smile was breaking, and mischief was afoot. She grew both to love and to dread it, and as I saw it in my own son I simply sighed, wondering silently about aquariums and dogs.

I hope some day to take Joshua to a concert featuring my friend's efforts at putting *The Inferno* to music. Knowing her talent, she will make it interesting, beautiful, dramatic, and powerful. And I hope Josh will understand the thought that goes into the design of this music, even though he may not fully comprehend why each instrument was used in a given measure, or how specific notes when pressed into symphonic service create emotional impact. This hope is a bit presumptuous, if I think about it, for I will *never* comprehend how Joshua will hear this piece of music, or any other work of art he will encounter. Will he isolate the twisting, turning nature of shimmering themes, alternating keys, pulsating rhythms? Or will he summarize the interaction of these hundreds of notes, and allow a synthetic musical monolith to pass into his awareness and inspire him? Even though I will never really know what he perceives, I anticipate that some day he will ask me what is the *right* way to hear a symphony. And I suppose I will have to teach him that there is no one right way to appreciate any work of art. When one encounters a mysterious masterpiece filled with unknown but thrilling complexities, one is not left with the answer, one is simply left with a point of view.

Further Reading

Chapter One

LeDoux, J. 1996. *The Emotional Brain: The Mysterious Underpinnings of Emotional Life.* p. 12. New York: Simon and Schuster

LeDoux, J. 1996. *ibid.* p. 268. New York: Simon and Schuster

Griffiths, P. 1997. *What Emotions Really Are.* p. 247. Chicago: The University of Chicago Press

Chapter Two

Blok, L. J., de Ruiter, P. E., Brinkmann, A. O. 1996. Androgen receptor phosphorylation. *Endocrine Research* **22**:197–219

Blum, D. 1997. *Sex on the Brain.* New York: Viking Penguin

Charlton, R. S., Quatman, T. 1997. A therapist's guide to the physiology of sexual response. In *Treating Sexual Disorders*, ed. I. D. Yalom. pp. 29–58. San Francisco: Jossey–Bass

Crenshaw, T. L., Goldberg, J. P. 1996. Basic principles of sexual pharmacology. In *Sexual Pharmacology.* p. 95. New York: W. W. Norton

Fletcher, P. C., Dolan, R. J., Frith, C. D. 1995. The functional anatomy of memory. *Experientia* **51**:1197–207

Katzenellenbogen, B. S. 1996. Estrogen receptors: bioactivities and interactions with cell signalling pathways. *Biology of Reproduction* **54**:287–293

Martini, L., Celotti, F., Mecangi, R. C. 1996. Testosterone and progesterone metabolism in the central nervous system: cellular localization and mechanism of control of enzymes involved. *Cellular and Molecular Neurobiology* **16**:271–82

Meisel, R. L., Sachs, B. D. 1994. The physiology of male sexual behavior. In *The Physiology of Reproduction*, ed. E. Knobil, J. D. Neill. pp. 3–106. Vol. 2. New York: Raven Press

Pfaff, D. W. *et al.* 1994. Cellular and molecular mechanisms of female reproductive behaviors. In *The Physiology of Reproduction*, ed. E. Knobil, J. D. Neill. pp. 107–220. Vol. 2. New York: Raven Press

Ramirez, V. D., Zheng, J. 1996. Membrane sex-steroid receptors in the brain. *Frontiers in Neuroendocrinology* **17**:402–39

Rissman, E. F. *et al.* 1997. Estrogen receptors are essential for female sexual receptivity. *Endocrinology* **138**:507–10

Chapter Three

Berridge, K. C. 1996. Food reward: brain substrates of wanting and liking. *Neuroscience and Biobehavioral Reviews* **20**:1–25

Drewnowski, A. 1997. Taste preferences and food intake. *Annual Review of Nutrition* **17**:237–57

Flier, J. S., Maratos-Flier, E. 1998. Obesity and the hypothalamus: novel peptides for new pathways. *Cell* **92**:437–40

Herman, P. 1996. Human eating: diagnosis and prognosis. *Neuroscience and Biobehavioral Reviews* **20**:107–11

Karhunen, L., Haffner, S., Lappalainen, R., Turpeinen, A., Miettinen, H., Uusitupa, M. 1997. Serum leptin and short-term regulation of eating in obese women. *Clinical Science* **92**:573–8

King, N. A., Tremblay, A., Blundell, J. E. 1997. Effects of exercise on appetite control: implications for energy balance. *Medicine and Science in Sports and Exercise* **29**:1076–89

Langhans, W. 1996. Role of the liver in the metabolic control of eating: what we know and what we don't know. *Neuroscience and Biobehavioral Reviews* **20**:145–53

Montague, C. T. *et al.* 1997. Congenital leptin deficiency is associated with severe early-onset obesity in humans. *Nature* **387**:903–8

Sakurai, T. *et al.* 1998. Orexins and orexin receptors: a family of hypothalamic neuropetides and G protein-coupled receptors that regulate feeding behavior. *Cell* **92**:573–85

Seeley, R. J. *et al.* 1997. Melanocortin receptors in leptin effects. *Nature* **390**:349

Spiegelman, B. M., Flier, J. S. 1996. Adipogenesis and obesity: rounding out the big picture. *Cell* **87**:377–89

Chapter Four

Adolphs, R., Tranel, D., Damasio, H., Damasio, A. R. 1995. Fear and the human amygdala. *Journal of Neuroscience* **15**:5879–91

American Psychiatric Association 1994. *Diagnostic and Statistical Manual of Mental Disorders*. Fourth Edition. Washington, D.C.: American Psychiatric Association

Baddely, A. 1992. Working memory. *Science* **255**:556–9

Diamond, D. M., Rose, G. 1994. Stress impairs LTP and hippocampal-dependent memory. *Annals of the New York Academy of Sciences* **746**:411–14

Goldman-Rakic, P. S. 1993. Working memory and the mind. In *Mind and Brain: Readings from Scientific American Magazine*, ed. W. H. Freeman. pp. 66–77. New York: Freeman

Halgren, E. 1992. Emotional neurophysiology of the amygdala within the context of human cognition. In *The Amygdala: Neurobiological Aspects of Emotion, Memory and Mental Dysfunction*, ed. J. Aggleton. pp. 191–228. New York: Wiley-Liss

LeDoux, J. E. 1995. Emotion: clues from the brain. *Annual Review of Psychology* **46**:209-235

McClelland, J. L., McNaughton, B. L., O'Reilly, R. C. 1995. Why there are complementary learning systems in the hippocampus and neocortex: insights from the successes and failures of connectionist models of learning and memory. *Psychological Review* **102**:419–57

Nikelly, A. G. 1992. The pleonexic personality: a new provisional personality disorder. *Journal of Adlerian Theory* **48**:253–60

Rosenzweig, M. 1996. Aspects of the search for neural mechanisms of memory. *Annual Review of Psychology* **47**:1–32

Chapter Five

Holden, C. 1997. Moods and sleep. *Science* **275**:1071–2

Kay, S. A. 1997. PAS, present and future: clues to the origins of circadian clocks. *Science* **276**:753–4

King, D. P. *et al* 1997. Positional cloning of the circadian clock gene. *Cell* **89**:641–53

Liu, C., Weaver, D. R., Strogatz, S. H., Reppert, S. M. 1997. Cellular construction of a circadian clock: period determination in the suprachiasmatic nucleus. *Cell* **91**:855–60

Oren, D. A., Terman, M. 1998. Tweaking the human circadian clock with light. *Science* **279**:333–4

Reppert, S. M., Weaver, D. R. 1997. Forward genetic approach strikes gold: cloning of a mammalian clock gene. *Cell* **89**:487–90

Sassone-Coral, P. 1997. PERpetuating the PASt. *Science* **389**:443–4

Sassone-Coral, P. 1998. Molecular clocks: mastering time by gene regulation. *Nature* **392**:871–4

Sawyer, L. A. 1997. Natural variation in a *Drosophila* clock gene and temperature compensation. *Science* **278**:2117–20

Shigeyoshi, Y. *et al*. 1997. Light-induced resetting of a mammalian circadian clock is associated with rapid induction of the mper1 transcript. *Cell* **91**:1043–53

Chapter Six

Archer, J. 1991. The influence of testosterone on human aggression. *British Journal of Psychology* **82**:1–28

Bock, G. R., Goode, J. A. 1996. *Genetics of Criminal and Antisocial Behavior*. London: Wiley

Cadoret, R. J. 1978. Psychopathology in adopted-away offspring of biological parents with antisocial behavior. *Archives of General Psychiatry* **35**: 1171–5

Cadoret, R. J., Cain, C. 1981 Environmental and genetic factors in predicting adolescent antisocial behavior. *The Psychiatric Journal of the University of Ottowa* **6**: 220–5

Coccaro, E. F., Bergeman, C. S., McClearn, G. E. 1993. Heritability of irritable impulsiveness: a study of twins reared together and apart. *Psychiatry Research* **48**:229–42

Daly, M., Wilson, M. 1994. Evolutionary psychology of male violence. In *Male Violence*, ed. J. Archer. London: Routledge

Hen, R. 1996. Mean genes. *Neuron* **16**:1–20

Higley, J. D. *et al* 1992. Cerebrospinal fluid monoamine and adrenal correlates of aggression in free-ranging Rhesus monkeys. *Archives of General Psychiatry* **49**:436–41

Lyons, M. J. *et al* 1995. Differential heritability of adult and juvenile antisocial traits. *Archives of General Psychiatry* **52**:906–15

Mednick, S. A., Gabrielli, W. F., Hutchings, B. 1984. Genetic factors in criminal behavior: evidence from an adoption cohort. *Science* **224**:891–3

Nelson, R. J. *et al* 1995. Behavioural abnormalities in male mice lacking neuronal nitric oxide synthase. *Nature* **378**:383–6

Raleigh, M. J. *et al* 1991. Serotonergic mechanisms promote dominance acquisition in adult male vervet monkeys. *Brain Research* **559**:181–90

Wilson, E. O. 1995. *On Human Nature*. Penguin Books, London.

Chapter Seven

Brucker, G. 1998. The formation of the Florentine dominion. In *Florence: The Golden Age, 1138–1737*. pp. 157–191. Berkeley: University of California Press

Clayton, P. J. 1994. Unipolar depression. In *The Molecular Basis of Psychiatry*, ed. G. Winokur, P. J. Clayton. Philadelphia: W.B. Saunders

Kahn, A. 1990. Suicide over the life cycle: risk factors, assessment and treatment of suicidal patients. In *Principles of Psychotherapy with Suicidal Patients*, ed. S. Blumenthal, D. Kupfer. pp. 441–67. Washington, D.C.: American Psychiatric Press, Inc.

Kendler, K. S. *et al* 1992. Major depression and generalized anxiety disorder: same genes (partly), different environments? *Archives of General Psychiatry* **49**:716–22

Medina, J. J. 1998. The human brain and depression. In *Depression: How it Happens, How it Heals*. pp. 31–46. Irvine, CA: CME, Inc. and New Harbinger Press

Nemeroff, C. B. 1996. The corticotropin-releasing factor (CRF) hypothesis of depression: new findings and new directions. *Molecular Psychiatry* 1:336–42

Nemeroff, C. B. 1998. The neurobiology of depression. *Scientific American* 278:42–9

Posner, M. I., Raichle, M. E. 1994. Past images. In *Images of Mind*. pp. 1–28. New York: Scientific American Library

Roy, A., Segal, N. L., Sarchiapone, M. 1995. Attempted suicide among living co-twins of twin suicide victims. *American Journal of Psychiatry* 152:1075–6

Smith, R. H., Parrott, W. G., Ozer, D., Moniz, A. 1994. Subjective injustice and inferiority as predictors of hostile and depressive feelings in envy. *Personality and Social Psychology Bulletin* 20:705–11

Chapter Eight

Abel, T., Martin, K. C., Bartsch, D., Kandel, E. R. 1998. Memory suppressor genes: inhibitory constraints on the storage of long-term memory. *Science* 279:338–41

Arber, S. *et al* 1998. Regulation of actin dynamics through phosphorylation of cofilin by LIM-kinase. *Nature* 393:805–9

Bailey, C. H., Bartsch, D., Kandel, E. R. 1996. Towards a molecular definition of long-term memory. *Proceedings of the National Academy of Sciences, USA* 93:12445–52

Eichenbaum, H. 1997. How does the brain organize memories? *Science* 277:330–2

Hooper, J., Teresi, D. 1986. The three-pound universe. In *The Three-Pound Universe*. p. 16. New York: The Putnam Publishing Group

Kempermann, G., Kuhn, H. G., Gage, F. H. 1998. More hippocampal neurons in adult mice living in an enriched environment. *Nature* 386:493–5

Lisman, J., Malenka, R. C., Nicoll, R. A., Malinow, R. 1997. Learning mechanisms: the case for CaM-KII. *Science* 276:2001–2

Schacter, D. L. 1998. Memory and awareness. *Science* 280:59–61

Sheldon, M. *et al* 1997. Scrambler and yotari disrupt the disabled gene and produce a reeler-like phenotype in mice. *Nature* 389:730–3

Silva, A. J., Paylor, R., Wehner, J. M., Tonegawa, S. 1992. Impaired spatial learning in alpha-calcium calmodulin kinase mutant mice. *Science* 257:206–11

Tully, T. 1997. Regulation of gene expression and its role in long-term memory and synaptic plasticity. *Proceedings of the National Academy of Sciences, USA* **94**:4239–41

Index

optic chiasm, 160
orexin, 98
orphan receptors, 93

parabolic surgery, 84
parasympathetic nervous system, 76
PAS region, 175
penis, 28, 29
  blood flow, 29
  glans, 29
per, 175
perfect blood, 59
Philip IV, 128
pituitary, 128
PKA, 315
Plato, 60, 61, 184
polarize
  definition, 17
Pope Martin IV, 78
popliteal region, 160
post-traumatic stress disorder (PTSD), 114
potassium, 17
pride, 274, 275
promoter, 168
proopiomelanocortin (POMC), 89, 90
protein
  definition, 42
Purgatory
  physical description, 20
Purkinje cell, 5

Rac, 298
Ravenna, 133
receptors
  AMPA, 310
  intracellular, 52
  leptin, 85
  signal transduction, 51
Reelin, 285
reflex arcs, 28
rigui, 182
RNA polymerase, 168
Robo, 290, 292
Ryle, Gilbert, 265

Salvani, Provenzano, 247
scrambler, 285
Selective Serotonin Reuptake Inhibitors (SSRIs), 262
sensory cortex, 126
serotonergic system, 260

serotonin, 141, 147, 219, 220, 223, 224, 227, 259, 260, 262, 263, 271
  and anxiety, 141
sexual arousal
  definiton, 25
"shade", 60
short-term buffers, 102, 103, 104, 124, 144, 146, 147, 187, 227, 269, 270, 330
skull trepanning, 33, 36
sloth, 152
Sodomites, 38
SSRI, 262
Statius, 23
stria terminalis, 205
super-males, 223
suprachiasmatic nucleus (SCN), 157, 158, 159, 160, 161, 180, 181, 183, 186, 189
sympathetic nervous system, 74
synapse, 134

temporal lobe, 126
teratoma, 14
testosterone
  and aggression, 206
  and castration, 207
  and female sexual arousal, 48
  and male sexual arousal, 47
thalamus, 118
The Divine Comedy
  history, 18
tim, 175
Tomyria, 303
transitional cortex, 127
Tree of Knowledge, 83
triglyceride, 81
twins, 198

Unguentum Armarium, 2, 4, 6

Valium, 138
vetromedial nucleus, 39
Virgil, 23
von Leibniz, Gottfried, 142

wc-1, 173
wc-2, 173
working memory, 146, 147
wrath, 191

York Minster, 25

Zadar, 133